Biomarkers in Food Chemical Risk Assessment

Biomarkers in Food Chemical Risk Assessment

Edited by

Helen M. Crews and A. Bryan Hanley

CSL Food Science Laboratory, Norwich Research Park, Colney, Norwich NR4 7UQ, UK

CSL, the Central Science Laboratory, is an executive agency of the Ministry of Agriculture, Fisheries and Food, UK.

THE ROYAL
SOCIETY OF
CHEMISTRY

The proceedings of the Symposium 'Biomarkers in Food Chemical Risk Assessment', held on 1–3 March 1995 at CSL Food Science Laboratory, Norwich Research Park, UK

Special Publication No. 175

ISBN 0-85404-790-5

A catalogue record for this book is available from the British Library

Published by The Royal Society of Chemistry,
Thomas Graham House, Science Park, Milton Road,
Cambridge CB4 4WF, UK

Printed by Hartnolls Ltd, Bodmin, Cornwall, UK

Foreword

David G. Lindsay

CHIEF SCIENTIST'S GROUP (FOOD), MINISTRY OF AGRICULTURE, FISHERIES AND FOOD, ROOM 219, NOBEL HOUSE, 17 SMITH SQUARE, LONDON, SW1P 3JR, UK

Up to the present time, risks to humans from chemical substances in the diet have mostly been assessed using methods of risk assessment which were developed for the pharmaceutical and agro-industrial chemical sector. These same methods have been used for chemicals that are added to food in its production. The assessment of risk is based on information obtained from feeding laboratory animals with the chemical substances, until toxicity is produced and, increasingly, the use of in vitro tests for toxicity. The application of safety factors to the dose at which there are no observed effects varies according to the reliability of the experimental data and, whether or not data are available in humans. This approach enables acceptable dietary intakes to be derived which provide a large margin of safety when the chemical is consumed.

The use of these methods has ensured that consumers have been protected against exposure to overtly toxic compounds but increasingly it has become evident that this methodology has severe limitations when applied generally to chemical substances which are associated with undertaking statistically valid dose-response lifetime studies in animals. This limits the current approach to those chemicals with a large potential market.

Other limitations include the use of high doses of the compound which, if present as a natural constituent of the diet, frequently have to be administered to animals to observe toxicity. These high doses cause nutritional and metabolic changes which do not occur at lower doses. At the usual levels of dietary exposure these effects would not be observed. In addition, dietary factors can modulate the metabolism of the compound.

As more attention is focused on chemicals naturally present in food, the traditional approaches used in the past become impracticable and frequently of questionable scientific validity. The resources invested in attempting to identify naturally occurring chemical risks in the food supply, particularly carcinogenic risks, have detracted resources from focusing on identifying naturally occurring chemicals which confer benefits.

Great interest is being taken in the natural constituents of food, not only for their potential for functionality in food production but also because of the increasing relevance of diet as an important factor in the aetiology of cancer and cardiovascular disease. The constituents of those commodities which exert this protective affect need to be identified and their mechanism of action evaluated. Epidemiological studies suggest that increased fruit and vegetable consumption can protect consumers against these diseases.

The concept of risk assessment for naturally occurring components of food would benefit from a radical new approach in order to enable health benefits to be delivered. Ideally the approach should be based on studying early events in the mechanisms which lead to disease, and monitoring the influence of various dietary constituents on these events. Sufficient is already known about the likely early metabolic events which could lead eventually to cancer and cardiovascular disease to make this feasible. Since all such events are likely to be influenced by genetic factors it becomes increasingly necessary to develop tools which enable these influences to be identified and where possible quantified.

The great interest currently being shown in the development of biomarkers, which is reflected in this workshop, has occurred not only as a result of the need to find new approaches to the evaluation of food safety, but also because developments in techniques both of analytical chemistry and molecular biology, have now made it feasible to undertake human biomonitoring studies. Biomarkers can be developed to measure intake, exposure, dose-response and effect. With cancer and coronary heart disease as the targets of greatest interest, the most pressing needs are to study the effects of oxidative reactions and the factors which influence their outcome on the damage to specific cellular DNA, lipids and proteins. For most biomonitoring studies it will require the use of feasible, non-invasive, methods which can be validated (initially in animal systems) as true measures of the effects on target organs, and which are capable of measuring dose-responses which will enable risks and benefits to be assessed.

Preface

The importance of diet as a contributing factor in human disease has been extensively acknowledged, frequently cited, repeatedly reported but rarely quantified. One major reason for this apparent lack of a definable role for dietary factors is the difficulty in characterising those processes which lie on the pathway between ingestion and a measurable effect on health. Biomarkers have been used with varying degrees of success to assess health effects of environmental factors such as workplace exposure to noxious chemicals. In occupational studies the link between cause and outcome is relatively easily established using biomarkers of exposure and effect. In contrast, for food chemical risk assessment there is a paucity of information available on the nature of likely biomarkers of effect or even, in many cases, comprehensive measurements of exposure. In addition, exposure to food chemicals is, in the vast majority of cases, chronic and at low levels which does not lead inexorably to a single biological outcome. One solution to this problem is the development and better understanding of biomarkers to assess the impact of diet on health and disease. While it may not be possible to establish a quantitative link between certain dietary factors and specific disease processes, the development of representative biomarkers can help to link research in epidemiology, nutrition, chemistry, cell biology, risk assessment and toxicology and enable it to be targeted towards a common goal - a better multidisciplinary understanding of food chemical risk assessment.

The meeting was organised by the CSL Food Science Laboratory, Norwich Research Park in March 1995 and marked the first anniversary of the setting up of the enlarged Central Science Laboratory as an executive Agency of the Ministry of Agriculture, Fisheries and Food (MAFF). It is particularly appropriate that a significant theme of the meeting was the role of biomarkers in risk assessment and assessment of their potential value to MAFF in particular and to other similar organisations. Efforts to determine the relative risks to humans from dietary factors must on the one hand be underpinned by excellent and innovative science and must also strive to provide coherent and useful answers. It is important to ask relevant questions even if the answers are not always clear and obvious. These considerations were to the fore throughout this meeting and we would like to applaud the efforts of all the contributors who did so much to stimulate thought and discussions.

The organisation of any meeting is a team effort. We are grateful to the Scientific Committee for their input and enthusiasm and to the Organising Group at CSL, Norwich for their efforts before, during and after the meeting - particularly on the last day when the venue for the lectures had to be changed at the last minute. We would like to thank the Institute of Food Research for hosting most of the scientific sessions. Finally, a multidisciplinary meeting such as this is particularly dependent upon the participants and their enthusiasm and we would like to thank all those who attended and we look forward to the next Norwich Biomarkers Meeting.

H.M. Crews
A.B. Hanley
September 1995

Contents

Key Issues in the Use of Biomarkers for Assessing Risks from Food Chemicals

J. C. Sherlock

HEAD OF FOOD SCIENCE DIVISION I, MINISTRY OF AGRICULTURE, FISHERIES AND FOOD,
ERGON HOUSE C/O NOBEL HOUSE, 17 SMITH SQUARE, LONDON SW1P 3JR, UK

The purpose of this paper is to give a general over-view of some of the important issues in the field of biomarkers. Biomarkers is very much a buzz word these days, however, it is never entirely clear just what is meant by "biomarker"; for example it might refer to the presence of DNA adducts as biomarkers of exposure to some carcinogen or other, it might also refer to some specific type of cell damage as a biomarker, or perhaps it might refer to neurological signs and symptoms as biomarkers.

The simplest definition of biomarker would be "a measurement or observation on the whole or part of an organism which provides an indication of exposure to some substance or group of substances". This would encompass all organisms and substances which range from microbes and viruses through to environmental contaminants, such as PCBs or mercury. A trivial example could be a high blood pressure reading being a biomarker for exposure of man to an unhealthy diet or to some external stimulus.

This type of definition does not go quite far enough in the present climate: it needs to be enhanced by incorporating into it some mention of "risk". Therefore the addition of a sentence such as "and which gives an indication of the risks to health associated with that exposure".

Given that the focus of this collection of papers is on man and food chemical risk assessment, biomarker could be defined as:

a measurement made on body tissue, body fluid or excretion to give a quantitative indication of exposure to a chemical and which may give an estimate of the risks consequent on that exposure.

Lest it be thought that a new science has been invented rather than a new word it is as well to consider which biomarkers have been used in days gone by and which are still in regular use.

Some 500 years ago Paracelsus[1] laid down part of the basis for risk assessment, he was primarily concerned with intake or exposure when he said "all substances are poisons: there is none which is not a poison. The right dose differentiates a poison and a remedy" and he was only putting into words a common sense used by people who have poisoned others since time began - namely the bigger the dose, or intake, the greater the risk and the more dramatic the effect.

In the previous 2 decades considerable attention has been focused on lead in man with extensive studies of exposure, the origins of this exposure, the accumulation of lead in man, and the risks from exposure to lead. Biomarkers for lead exposure have included lead in blood[2], lead in urine[3], lead in teeth[4], free erythrocyte protoporphyrin[5], d-aminolaevulinic acid dehydratase[6], lead in hair[7] and lead in nails[8]. Some of the methods directly measure lead itself whilst others measure things upon which lead is thought to have an effect such as enzymes. The uncertainty in the work on lead, apart from of those obviously associated with analysis, was always and still is the degree of risk. Few people doubt that lead is harmful, indeed quite modest elevations in blood lead concentrations are sufficient cause to have workers removed from the source of exposure. Inhibition of enzyme systems is undesirable but whether partial inhibition carries with it a serious risk of damage to health is another matter. For many years the argument has been that exposure of young children to low levels of lead may result in some impairment of mental development and a consequent deficit in expected IQ[9]. Though the deficit to individuals may only be small the accumulated effect nationwide is much more serious.

The utility of biomarkers can be illustrated by reference to lead exposure. The toxicological experts can make a judgement about the correspondence between the magnitude of the biomarker, in this case the concentration of lead in blood, and the health effect, be it IQ detriment or enzyme inhibition. On this basis a maximum acceptable or tolerable value for the biomarker can be determined. The acceptable or tolerable level of a biomarker can often be correlated with some other measurement, in this case lead in air, lead in water or lead in food, and this correlation used to set a standard for the exposure medium.

Expressed mathematically the above statement reduces to:

$$B = \text{concentration or level of biomarker for substance}$$
$$C = \text{concentration of substance in exposure medium}$$
$$B = F(C) \text{ relationship between biomarker and concentration of substance}$$
$$\text{whence } C = F^{-1}(B)$$
$$\text{if } B_T = \text{maximum tolerable level or concentration of biomarker}$$
$$\text{then } C_T = F^{-1}(B_T)$$
$$\text{where } C_T = \text{inferred maximum tolerable concentration of substance in exposure medium}$$

For the case of lead in water, the relationship between lead in blood concentration, PbB, and lead in tap water concentrations, PbW, has been established[10] for bottle-fed infants to be:

$$PbB \ (\mu/dl) = 5.5 + 3.3 \ ^{3}\sqrt{PbW(\mu g/l)}$$

Assuming B_T (PbB) $= 15\mu g/dl$ (say)

then C_T (PbW) $= \left[\dfrac{15-5.5}{3.3}\right]^{3} = 24\mu g/l$ *

* for the average infant

In much the same fashion measurements of arsenic in hair or urine[11,12] have been used as biomarkers for arsenic exposure be it accidental, deliberate or occupational. Although in this instance whilst measurement of arsenic in urine will act as a biomarker of exposure to both inorganic arsenic, which is probably carcinogenic[13], and organo-arsenic from fish, which is not toxic at normal exposure levels, accumulation of arsenic in hair is indicative of exposure to inorganic arsenic compounds only. Measurement of lipid soluble organic chemicals in the fat component of breast milk will give some indication of past exposure to these contaminants although for obvious reasons this measurement is useful only on women. Dioxins, PCBs and other chlorinated organic compounds are obvious analytes.

Perhaps the most familiar biomarker of all is one that might not spring to mind in the context of this collection of papers. The biomarker that most people are familiar with, although one hopes not through practical experience, is the breathalyser. The breathalyser provides a simple measurement of oxidisable alcohol-related organic material in breath which is related to the amount of alcohol in blood. After allowing for the time between the cessation of drinking and the taking of a measurement, the concentration of oxidisable material in breath will give a reasonable estimate of the amount of alcohol consumed or perhaps "of the exposure to alcohol".

The question arises why the current interest in biomarkers and why the need for research and method development. At its simplest the need arises because, in the field of food chemicals, one of the most complex and intractable problems is estimating what the exposure to them might be. It is simple enough to make back of the envelope calculations but these can go badly wrong particularly because, even though science is sophisticated, scientists sometimes tend to forget that there are things we do not know and knowledge will remain imperfect. Then again there may be others who have an interest in colouring the picture blacker or whiter than the facts would indicate by making gross overestimates or deliberate underestimates of exposure. One of the difficulties regulators face is achieving the right balance between their natural caution in wishing to err on the safe side thus affording consumers maximum protection and the need to be sensible and reasonable so as not to burden industry with costs arising from action which is not warranted. At present most of the estimates of exposure to food additives are made by calculation rather than by observation. Because of the need for sound defensible science it is essential that things are thought through carefully; this need is illustrated by reference to the case of Annatto.

When the EC Colours Directive was being negotiated it became apparent that the proposed levels of use of Annatto were too low for industry yet calculations indicated that at those levels consumers would exceed the acceptable daily intake (ADI) by a substantial amount. In other words it seemed as if UK industry would be getting the worst of both worlds, permitted concentrations of Annatto would not be high enough to achieve the desired effect and even then intakes would be too high. By gathering together detailed consumption data for individual coloured foods and combining this with accurate information on levels of use of Annatto by industry a more precise but still conservative estimate of the intake of Annatto was made. This revised intake was well within the ADI and so the European Commission was able to accommodate UK needs. The revised intake estimate for Annatto is still erring on the conservative side but it is not known by how much.

The three European Union Directives on sweeteners[14], on colours[15] and on additives other than sweeteners and colours[16] each require monitoring of the consumption of the additives they control. Estimates of average intakes can be produced by gathering together information on industry usage of the additives and simply dividing by the number of people living in the UK[17]. However, average intakes whilst useful say little about the exposure of individuals and this is where biomarkers could have a vital role to play.

For some additives, such as artificial sweeteners and colours, it should be possible to find biomarkers which will at least give an indication of exposure. Work currently underway at the Central Science Laboratory (CSL) of Norwich is investigating urinary excretion of sweeteners and colours. Judicious use of biomarkers to study additive intake should demonstrate in a cost-effective fashion that intakes of additives are not simply less than the ADI but probably several orders of magnitude less than the ADI. There would be little purpose in using this methodology widely because it will be expensive to apply, however, there would seem to be a case for applying the methodology to those additives for which intakes are estimated to be nearest to their respective ADIs and thus a **potential** cause of consumer concern even though few additives fall into this category. Some might argue that consumer concerns are not justified because even a life times consumption of an additive at the ADI every day is judged to be without appreciable risk. However consumer concerns are genuine and it is right that they be addressed.

Use of biomarkers to estimate additive intake should be relatively straightforward, but there is a need to think about how the results will be used and interpreted before undertaking the studies. For example measurements of the daily urinary excretion of an additive could give a very misleading picture of the long-term exposure to the additive. Figure 1a shows hypothetical data for a normal distributed daily intake of an additive, the mean being 12 units/day with a standard deviation of 4 units/day. Figure 1b shows the **same data** but with intakes averaged over a week. The pictures presented by the two figures are different. If the ADI for the additive was 14 units/day, then in Figure 1a the ADI would be exceeded by more than 30% of observations but in Figure 1b it would be exceeded by about 5% of observations. It is Figure 1b which gives the more useful estimate of exposure, day-to-day variations in intake are not as important as the long-term picture when dealing with low level exposure to substances at concentrations well below those where there is even a significant chance of chronic effects.

Whilst it would be comforting to think that calculated estimates of exposure are always unduly conservative, that is they are overestimates, it is unlikely that this is always the case. Work on biomarkers undertaken at CSL[18] provided useful information on actual exposures to di-2-ethylhexyl adipate (DEHA) compared with theoretical estimates. DEHA is a plasticizer used in plasticized PVC film and can enter the diet as a result of consumption of food which has been wrapped in cling film at some time. There are of course other sources of exposure to DEHA such as its use in lubricants and textiles. The study at CSL investigated the urinary excretion of

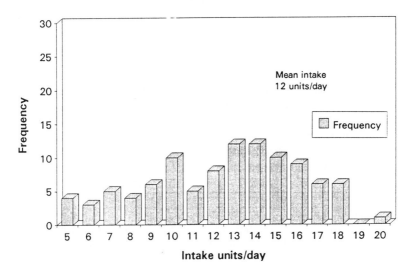

Figure 1a: Intake measured each day

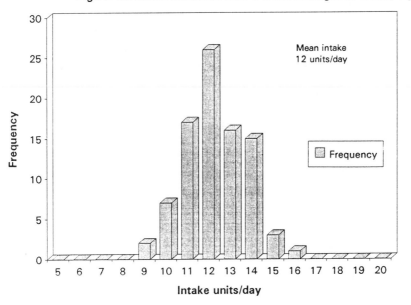

Figure 1b: Intake measured over a week averaged to a daily figure

metabolites of DEHA. Calculated estimates of the maximum likely intake of DEHA indicated[19] that these were unlikely to exceed about 8mg/day. The study of the excretion of DEHA in urine from 112 individuals found that actual intakes range from <1mg/day up to 11mg/day with only 3 individuals exceeding 8mg/day. Bearing in mind the rough and ready nature of the calculated intake and the difficulties in conducting the analytical experimental study, the concordance between the two estimates of maximum likely intake is remarkable.

It should be possible to extend use of biomarker techniques to study the exposure of individuals to other contaminants and thus provide reliable information on likely intakes. The difficulty with this approach is that whilst biomarker studies will give a measure of total exposure, it is unlikely that the exposure from food alone would be identified excepting in special cases where food is the only source of exposure. This weakness will not cause a major problem if food is the main source of exposure or where measurements are being made to produce an approximate intake. However, if regulatory action may follow from the interpretation of the results of the measurements, then caution must be exercised in that interpretation.

One of the most difficult questions facing advisors on the safety issues is how to judge the risks to human health arising from low level exposure to carcinogens. These difficulties extend beyond those concerned with food safety and include exposure to environmental carcinogens through inhalation or through water and from exposure to carcinogens which may be found in industry. In the case of exposure to ionising radiation from radioactivity matters are apparently straightforward because, it is thought, there is a well-defined linear relationship between the excess risks of developing a fatal cancer and the dose of radiation received; this applies only to relatively low doses such as those received by workers in industry or indeed the general population. In the case of chemical carcinogens matters are far from simple. The uncertainties in exposure to radioactivity estimation pale into insignificance compared with the uncertainties in estimating the risks to health arising from chemical exposures. This means that it is difficult if not impossible to judge the priorities for action to reduce human exposure to low levels of carcinogens or suspected carcinogens or indeed whether society need bother reduce exposure. Assuming that resources are limited, which they always are, and that government or industry must spend its resources as wisely as possible where they will do most good, then where is action to be taken in respect of the following carcinogens or suspect carcinogens?

- nitrosamines
- polynuclear aromatic hydrocarbons
- aflatoxins
- ochratoxin
- patulin
- ethyl carbamate
- benzene
- furfural
- coumarin
- dioxin

All of these may produce cancers in animals, or indeed even in man, provided the dose is high enough or prolonged enough but consumers are exposed to very low levels of these compounds via the food chain. The risk from exposure to these compounds will depend on many different factors including dose, individual susceptibility, repair mechanisms, the presence in the diet of protective factors, perhaps even the age of the exposed individual.

Progress on biomarkers of exposure to carcinogens, such as

> measures of DNA damage
> DNA adducts of genocarcinogens
> haemoglobin adducts
> oncogene activation
> levels of carcinogen in body fluids

will take us perhaps one or two steps further in estimating risk but will not in the short-term stand the remotest chance of producing a quantitative estimate of risk, that for the time being must remain like the philosopher's stone. More work would be needed on the development from DNA damage through to tumour formation, as would work on repair mechanisms, variation of susceptibility from individual to individual and a host of other issues. At this stage even to know whether a given level of DNA adduct carries with it a serious risk of cancer would be helpful particularly if this is then combined with a large number of prospective measurements of specific adducts, say to aflatoxin or nitrosamines, in the UK population.

In summary biomarkers appear to be very much flavour of the month and they have much to offer those responsible for estimating risks be they from food chemicals or other sources. Use of biomarkers to judge risks or to set standards is not new. What science is doing now is beginning to probe areas which had previously been too hard to do. The potential for progress is there but it is going to take a long time, probably a decade at least, and a considerable investment of research money before scientists can truly begin to quantify risks.

References

1.   T.B. Paracelsus, Practica D. Theophrastic Paracelsi, quoted in *'Environment and Man'*, 1977, **6**, Blackie, Glasgow.
2.   H T Delves, J C Sherlock, M J Quinn, *Human Toxicol.*, 1984, **3**, 279.
3.   G W Monier-Williams, 'Trace Elements in Food', Chapman and Hall, 1950.
4.   H Needleman *et al*, *New Engl. J. Mid.*, 1979, **300**, 689.
5.   A F Labbe, *Clin. Chem.*, 1977, **23**, 256.
6.   P A Meredith *et al*, *Toxicology*, 1978, **9**, 1.
7.   R Gibson, *J. Human Nutr.*, 1980, **34**, 405.
8.   H C Hopps, *Sci. Tot. Environ.*, 1977, **7**, 71.
9.   M A Smith *et al*, 'Lead Exposure and Child Development: An International Assessment', Kluwer Academic, London, 1989.
10.   M J Quinn and J C Sherlock, *Food Add. and Contam.*, 1990, **7(3)**, 387.
11.   National Academy of Sciences, *Arsenic: Medical and Biologic Effects of Environmental Pollutants*, National Academy of Sciences, Washington DC, 1977.
12.   R M Brown *et al*, *Human and Exper. Toxicol.*, 1990, **9**, 41.
13.   W P Tseng *et al*, *J. Nat. Cancer Instit.*, 1968, **40**, 453.
14.   European Commission, *Directive on Sweeteners Used In Foodstuffs*, 94/35/EC.
15.   European Commission, *Directive on Colours Used in Foodstuffs*, 94/36/EC.
16.   European Commission, *Directive on Food Additives Other Than Sweeteners and Colours*, 95/2/EC.
17.   Ministry of Agriculture, Fisheries and Food, *Dietary Intake of Food Additives in the UK: Initial Surveillance*, Food Surveillance Paper No 37, HMSO, London, 1993.
18.   R Massey, *(IBID)*.
19.   Ministry of Agriculture, Fisheries and Food, *Plasticizers: Continuing Surveillance*, Food Surveillance Paper No 30, HMSO, London, 1990.

Approaches for Biomarker Studies

R. C. Massey

CSL FOOD SCIENCE LABORATORY, NORWICH RESEARCH PARK, COLNEY, NORWICH
NR4 7UQ, UK

1 INTRODUCTION

The use of biomarker-based techniques in occupation exposure is well established[1] but their application to food chemical risk assessment is comparatively new. Biomarkers have the potential to probe the sequence of steps in the aetiology of many chronic diseases and these have been described by the National Academy of Sciences and the National Research Council.[2] This model categorises biomarkers in terms of those reflecting: internal dose, biologically effective dose; early biological effect; altered structure/function; clinical disease and susceptibility. From the point of view of food chemical risk assessment biomarkers can be usefully divided into three broad categories - exposure, toxic effect and susceptibility. Biomarkers of exposure provide information on the degree of exposure of an individual to a particular food chemical. This takes the form of the amount of the chemical that is absorbed from the gastro-intestinal tract following ingestion and most often involves measurement of the parent chemical, metabolite or macromolecular adduct in urine or blood. Biomarkers of toxic effect reflect adverse biological responses to the chemical. In the case of biomarkers of susceptibility, information is provided about the inherent sensitivity of a particular individual to the chemical. The key feature of these biomarkers is that they are targeted at, and generate information on, individuals rather than the population as a whole.

Biomarkers may in principle be applied to the full spectrum of disease states in which food chemicals may have, at least in part, a causative role. These include neurological dysfunction, immunotoxicity, kidney disease, cancer and heart disease. As a consequence the variety of analytical techniques that are applicable is enormous ranging from the assessment of impairment of cognitive skills as a consequence of lead exposure to sophisticated mass spectrometric techniques for investigating and quantifying DNA adducts. Rather than attempt to review all of the analytical techniques that have been employed in biomarkers this paper will focus on some of the key areas with particular reference to the applications of biomarkers to the assessment of exposure and toxic effect.

2 BIOMARKERS OF EXPOSURE

2.1 General Considerations

Biomarkers of exposure have the potential to provide information on the actual amount of a food additive or contaminant that an individual is exposed to. Unlike other techniques such as total diet studies, duplicate diets, recall methods or record keeping[3] a biomarker approach addresses directly the issue of bioavailability, i.e. the extent to which the food chemical is absorbed from the gastro-intestinal tract. This can have a significant effect in terms of the proportion of the dietary dose that an individual is, in practice, exposed to. For instance, using radio-labelled lead Heard et al[4] have shown that whereas some 40 - 50% of lead given to fasting human volunteers is absorbed from the gut, uptake is only 7% when the metal is present as part of a meal. An additional refinement that is sometimes feasible is to examine the fraction of the absorbed dose that is associated with the target of biological damage. This has been employed to assess exposure to aflatoxin B_1 by measuring urinary excretion of the corresponding DNA adducts.[5] Such data provide information on exposure and may prove useful as an indicator of toxic effect although the biological significance of urinary DNA adducts has yet to be established.

Another advantage of a biomarker of exposure approach is that it provides direct information on the exposure of an individual to a particular food chemical. This is in contrast with another widely used technique, the total diet study,[6] in which food groups, representing the nation's average diet, are analysed thereby providing a mean intake figure for the food chemical in question. A biomarker approach enables the intake of representatives of particular sub-sets of the population to be probed including, for instance, potential "at risk" groups such as infants, pregnant women and extreme consumers.

Whilst studies employing biomarkers have significant advantages, both in terms of encompassing bioavailability and assessment of an individual's exposure to the food chemical, there are a number of potential drawbacks. If the aim of the study is to investigate food-derived exposure then care must be taken to ensure that the potential confounding influence of other exposure routes such as occupational, smoking, and the use of cosmetics and pharmaceuticals, is taken into account. Additionally, and particularly where the biomarker is a metabolite or adduct rather than the parent compound, the influence of inter-individual variability in the proportion of the compound converted into the metabolite/adduct being measured needs to be carefully examined during preliminary feasibility studies. In general insufficient attention has been paid to the effects of differing cytochrome P450 enzyme phenotypes on the reliability and accuracy of biomarkers of exposure.

A number of analytical techniques have been employed to quantify biomarkers of exposure and these are reviewed in detail elsewhere.[7] In cases where the biomarker involves the parent compound itself modification of the analytical method employed for measurement of the chemical in foods is often a sensible starting point. Thus graphite furnace atomic absorption spectroscopy has been used to quantify protein bound metal ions such as chromium in blood.[8] Due to its specificity and sensitivity GC/MS has often been employed, e.g. in the analysis of haemoglobin adducts of the cooked food mutagen and carcinogen MeIQx.[9] The technique also has the advantage of providing structural information on unknown compounds and, for instance, enabling identification of metabolites or adducts which may be used subsequently as biomarkers. ^{32}P-Postlabelling techniques have been widely used in the detection of DNA adducts[10] including aromatic hydrocarbons, aromatic amines and, with poorer detection limits, alkylating agents. In essence DNA is enzymatically digested to the deoxyribonucleoside-3'-monophosphates of normal and adducted nucleosides. The 5' position is then labelled with ^{32}P-ATP using polynucleotide kinase and the nucleotides separated by TLC and detected by autoradiography. A number of modifications of the procedure have been developed[11, 12] which permit excellent detection limits of up to 1 adduct in 10^{10} nucleotides. Postlabelling also has the advantage of detecting the presence of unknown adducts although, due to variability in labelling efficiency, absolute quantification of such adducts is not feasible and, in addition, other approaches would be needed to identify them. Immunological-based techniques, such as ELISA methods and radioimmunoassays, have the advantage over postlabelling of being easily automated, thereby enabling large numbers of samples to be analysed. They are however less sensitive for determining DNA adducts and, for example, Dunn[13] estimates that detection limits for immunoassays are up to 10 times poorer than postlabelling methods and require 100 times the amount of DNA.

2.2 Some Applications of Biomarkers of Exposure to Food Chemicals

2.2.1. Nitrate. Ingestion of nitrate may arise from its presence in drinking water and foodstuffs, particularly certain vegetables such as lettuce.[14] Nitrate is also employed as a preservative, together with nitrite, in cured meats such as bacon and salami. Recent concern over human exposure stems from the possibility of methaemoglobinaemia in infants and the suggestion that nitrate may be implicated in human cancer.[15] About 5% of the ingested nitrate in humans is microbially reduced to nitrite which may then react with secondary amines and similar chemicals with the resulting formation of genotoxic N-nitroso compounds.

In order to investigate exposure to nitrate Packer and co-workers[16, 17] have devised a biomarker approach based on the urinary excretion of residual nitrate. Analysis of the nitrate content of urine was achieved using an autoanalyser whereby nitrate was microbially reduced by E. coli and the resulting nitrite determined colorimetrically following reaction with sulphanilic acid and N-(-1-naphthyl)-ethylene diamine dihydrochloride. Measurement of 24 hour urine samples revealed that the amount of nitrate ingested in the preceding 24 hours was described by the term $[(N_u - 0.22)/0.55]$; where N_u is the amount of nitrate in the 24 hour urine sample, the 0.22 figure corrects for urinary excretion of nitrate derived from endogenous synthesis of nitrate and slow clearance of body pools and the 0.55 factor correcting for the fact that, on average, 55% of dietary nitrate is excreted in the urine within 24 hours. The method was not found to be suitable for those suffering from achlorhydria as the resulting bacterial colonisation of the stomach, and subsequent microbial reduction of nitrate, decreased the percentage of nitrate excreted in urine. Application of this methodology in a study involving over 300 subjects residing in 7 geographically distinct regions of the UK revealed the average daily intake of nitrate to be 157 mg. It is of interest to compare this figure with the estimate of intake derived from total diet studies, i.e. 54 mg per day.[14] This is clearly substantially lower than that provided by the urinary biomarker approach. The difference may, in part, be due to the fact that total diet studies do not include food eaten outside the home, nor the contribution from tap water or alcoholic beverages. Nevertheless the magnitude of the discrepancy is surprisingly large and further studies are needed, including assessment of the potentially confounding influence of nitrate formation from dietary protein.

2.2.2 Di-2-(Ethylhexyl) Adipate (DEHA). DEHA is widely employed as a plasticiser in PVC film, such film being used for food wrap purposes both in the home and in the packaging of retail foods. Concern over human exposure to DEHA results from findings of carcinogenicity in rodent feeding studies.[18] Exposure to DEHA will be heavily influenced by the extent to which an individual uses plasticised PVC film and also the manner in which it is used. For instance, higher levels of migration are known to occur if plasticised PVC film is used to wrap fatty foods.[19] In such circumstances a biomarker approach is potentially likely to be particularly well suited to assessing individual exposure.

To examine this Loftus *et al*[20] orally administered [2]H-labelled DEHA in gelatine capsules to human volunteers. Following analysis by derivatisation GC/MS, they reported that the parent compound could not be detected in either plasma or urine. 2-Ethylhexanoic acid (EHA) was found to be the major metabolite, with urinary levels accounting on average for 8.6% of the administered dose, suggesting that it might be a useful biomarker of exposure. To check for the potential influence of confounding factors a literature search was conducted to investigate non plasticised PVC film sources of exposure to EHA and its esters. Apart from use in some wood preservatives and floor coverings, where occupational exposure might well have a confounding influence, there was little evidence to suggest significant sources of exposure to EHA other than via plasticised PVC film. In a follow up study[21] a similar 24 hour urinary excretion of EHA was observed when DEHA was ingested in the form of a meal comprising sandwiches containing ham and cheese, the latter having previously been wrapped in plasticised PVC film.   The biomarker approach was then employed in a limited population survey involving 112 volunteers in 5 geographically distinct locations in the UK. This revealed that daily intakes of DEHA ranged from 0.5 to 10.5 mg with a median of 2.7 mg. These data are reasonably consistent with the maximum daily intake of 8.2 mg calculated from food consumption tables and the average level of DEHA detected in foods.[19]

2.2.3 Aflatoxins. The acute and chronic toxicity of aflatoxins has been established for some time[22] and aflatoxin B_1 (AFB1) is amongst the most potent carcinogens known. Aflatoxins have been found in a wide variety of foodstuffs including peanuts, pistachio nuts, maize and dried fruits, particularly in countries where climate or poor storage conditions favour the growth of moulds.[23]

Probably more work has been performed using biomarkers to investigate exposure to aflatoxins than any other food chemical.   There are numerous reports of free and hydroxylated aflatoxins, in particular the oxidative metabolite aflatoxin M_1 (AFM1), in urine.[24] A range of analytical techniques have been employed including TLC, HPLC and ELISA for end determination together with conventional sample extraction and clean up techniques and also immunoaffinity clean up columns.[25] In an extensive investigation by Zhu *et al*[26] some 252 urine samples were collected from 32 households in the Guangxi Autonomous Region of the People's Republic of China. A good correlation (r = 0.65) was observed between the total dietary intake of AFB1 and total urinary excretion of AFM1, with between 1 - 2% of the dietary AFB1 being present in urine as AFM1. These findings have subsequently been confirmed by Groopman *et al*[5]. AFM1 may also be detected in milk and Zarba *et al*[27] have estimated that between 0.09 to 0.43% of dietary aflatoxin is excreted as AFM1 in the milk of lactating women. Of potential wider applicability as a biomarker of exposure is the finding that AFB1 can be detected in blood following consumption of foods containing this mycotoxin.[28] However given the rapid clearance of AFB1 from blood by the liver (half life c.a. 30 minutes) this approach is unlikely to be of general utility.

Exposure to aflatoxins may also be estimated from information regarding adduct formation on macromolecules. The major AFB1-DNA adduct was identified by Essigmann *et al*[29] as 2,3-dihydro-2-(N[7]-guanyl)-3-hydroxy-AFB$_1$ (AFB1-N[7]-guanine). Excretion studies performed in the rat[30] revealed a highly significant correlation (r = 0.99) between AFB1 dose and 24 hour urinary excretion of AFB1-N[7]-guanine. In addition to DNA adducts the product of interaction between AFB1 and serum albumin has proved useful as a biomarker of exposure. Protein adducts are also likely to have advantages over DNA adducts in view of their longer half lives and greater concentrations. The major AFB1-serum albumin adduct, resulting from reaction of the lysine side chain amino group with the 8,9 epoxide of AFB1, has been identified as an amino ketone.[31] Studies in the rat have shown that albumin is the only serum protein that binds to AFB1 and that between 1 to 3% of a dose of AFB1 is bound to the protein in 24 hours.[32]

Human studies have confirmed the value of both the AFB1-N[7]-guanine and serum albumin adducts as biomarkers of dietary exposure to AFB1. Groopman *et al*[5] examined the relationship between dietary intake of AFB1 and urinary excretion of free AFB1, metabolites and the AFB1-N[7]-guanine adduct in a study involving 42 volunteers in the Guangxi Autonomous Region, People's Republic of China, an area of high risk of liver cancer. Dietary intakes were measured by TLC whilst two separate methods, HPLC/UV and competitive radioimmunoassay, were employed for the urine analysis. A highly significant correlation, r = 0.65, was observed between the 24 hour urinary excretion of AFB1-N[7]-guanine, as measured by HPLC/UV, and the dietary intake of AFB1 on the previous day. In contrast the correlation, r = 0.26, between dietary intake of AFB1 and the response from the radioimmunoassay was not significant at the 0.05 level. Some insight to this was provided by the HPLC/UV analysis which revealed that aflatoxin P$_1$ (AFP1) was a major metabolite in the urine but that its concentration was not dose related to dietary exposure of AFB1. As the monoclonal antibody used in the radioimmunoassay had a high recognition for AFP1 it is likely that this may, at least in part, explain the lack of association between AFB1 intake and response from the immunoassay. Samples from this study were also analysed by Gan and co-workers[33] to examine whether the serum albumin-AFB1 adduct could be used as a biomarker of exposure to AFB1. Serum samples were cleaned up by Sepharose and the serum albumin-containing fraction hydrolysed with Pronase and the digest further cleaned up with an immunoaffinity column prior to quantification using a monoclonal antibody competitive radioimmunoassay. A highly significant correlation, r = 0.69, was observed between dietary intake of AFB1 and the serum albumin AFB1 adduct. Similar high correlations between dietary exposure to AFB1 and urinary excretion of AFB1-N[7]-guanine[34] and serum albumin AFB1 adduct[35] have also been detected in studies carried out in The Gambia, another high risk area for liver cancer.

3 BIOMARKERS OF TOXIC EFFECT

3.1 General Considerations

Whilst biomarkers of exposure provide valuable information on the dose of a food chemical that an individual is exposed to they do not, of themselves, enable an assessment of the risks associated with that exposure to be made. What is required is a parameter which changes in a measurable manner as a result of exposure and reflects the consequences of that exposure. The progression of most diseases is multi-stage and in principle it may be possible to identify biomarkers of toxic effect for each stage. From the view point of enabling preventative action to be taken biomarkers for the earlier stages are to be preferred, i.e. before the disease has progressed too far. Once such putative biomarkers have been identified their value as indicators of risk may be assessed in animal studies, human prospective cohort studies and epidemiological investigations.

The science of biomarkers of toxic effect is still very much in its infancy. One of the major problems is that the detailed mechanism of the progression of most diseases is only partially understood. As a consequence selection of biomarkers can only be made from an incomplete database. In addition prospective human studies to assess the usefulness of biomarkers are of necessity long term and expensive to perform. Whilst cohort studies can be designed with regard to smoking or occupational exposure this is less simple for food chemicals as everybody eats food and the exposure levels are lower than in the case of smoking or occupational exposure. A biomarker of toxic effect may in principle be specific to the compound of interest or non specific e.g. chromosomal aberrations or DNA strand breaks. To facilitate assessment of the risks associated with a particular food chemical specific biomarkers of toxic effect are to be preferred although at present non specific biomarkers are probably a more realistic target. In this context it is worth noting that whilst DNA adducts can be specific in relation to exposure, as in the case of AFB1-N^7-guanine, they do not, as yet, provide a quantitative assessment of the risks associated with the subsequent biological consequences.

In general biomarkers of toxic effect are likely to involve cellular, biochemical or molecular alterations.[36] As a consequence analytical techniques will tend to focus on cellular and molecular biological techniques, in contrast to methods employed for the measurement of chemicals in food. Analytical procedures potentially suitable for biomarkers of toxic effect are reviewed in detail by Ungar *et al.*[7]

A number of methods exist for measurement of DNA strand breakage including the recently developed COMET assay.[37] Unscheduled DNA synthesis is useful for assessing damage to DNA that is repairable by new DNA synthesis. It involves measurement of the rate of incorporation of nucleotides into newly synthesised DNA[38] using ^{3}H-labelled thymidine and quantification by autoradiography or liquid scintillation counting. One drawback with the technique with regard to its use as a biomarker of toxic effect is that adducts that are not excised, and which may therefore be the more relevant to subsequent cancer risk, will not be detected. Mutagenesis detection has considerable potential as a biomarker of toxic effect as mutations are likely to be persistent in contrast to DNA adducts. A number of mutations may have little biological consequence and their presence simply reflects exposure to mutagens. However, mutations occurring at certain specific sites within key genes, such as proto oncogenes, are likely to have serious biological consequences and these may be suitable as biomarkers of toxic effect. For instance, selected amino acid substitutions in the p21 protein product of the *ras* oncogene enable it to effect cell transformations and immunological techniques, using monoclonal antibodies, may be used to detect mutant forms of the protein.[39] Alternatively, mutations in the DNA may be detected, using the recently developed polymerase chain reaction technique, to amplify the target DNA prior to sequencing, and this has recently been employed in identification of specific mutations in the p53 tumour suppressor gene.[40]

Cytogenetic tests are widely used in the measurement of toxicity.[41] Proto-oncogenes may be activated by chromosomal rearrangements[42] and such alterations may also inactivate tumour suppressor genes.[43] As a consequence it is feasible that cytogenetic assays may have potential as biomarkers of toxic effect. Chromosomal aberration analysis involves measurement of alterations, such as gaps, breaks and fragments in one or both chromatid, with visualisation of the damage by staining and subsequent microscopic counting.[44] Sister chromatid exchange, i.e. symmetrical exchanges between sister chromatids, is also a commonly employed test and utilises differential staining with the thymidine analogue 5-bromo-2'-deoxyuridine. Another widely used cytogenetic test involves the detection of micronuclei, i.e. small additional nuclei formed by exclusion of chromosome fragments or whole chromosomes lagging at mitosis.[41]

3.2 Some Potential Candidate Biomarkers of Toxic Effect

3.2.1 Cytogenetic Alterations. To test the usefulness of cytogenetic rearrangements as biomarkers of toxic effect a large prospective cohort study is being conducted in the Nordic countries.[45] This has been running since 1970 and involves over 3000 subjects, largely selected on the basis of occupational exposure. To assess the predictive value for subsequent cancer risk peripheral blood lymphocytes from each subject were examined, on a single occasion, during the period 1970 and 1988 for one or more of the following: chromosome aberrations, sister chromatid exchange and micronuclei. Subjects with cancer diagnosed before the cytogenetic assay were excluded from the study. To minimise the effects of inter-laboratory variation individual results from each of the nine laboratories involved were classified, for each of the 3 tests, as low (1-33 percentile, of each laboratory), medium (34-66 percentile) or high (67-100 percentile). During the period 1970 to 1991 a total of 85 incidences of cancer had been diagnosed in the cohort. Of these 66 had been previously assessed for chromosome aberration and there was a statistically significant linear trend between the magnitude of chromosome aberration and occurrence of cancer. 39 cancers were detected in the high chromosome aberration group compared with an expected figure of 18.9 derived from national cancer registry statistics. 49 of the cancer cases had been monitored for sister chromatid exchange prior to cancer diagnosis but no association between the results of this test and cancer incidence was apparent. Of the subjects screened for micronuclei 11 cancer cases were diagnosed but again no association was apparent. Taken overall these interim results from the Nordic Study Group suggest that chromosome aberrations appear to be a relevant biomarker of toxic effect for occupational exposure. The indications are that sister chromatid exchange and micronuclei may be less useful but more data are required before firm conclusions can be drawn. Whilst this study represents a major advance it is not clear whether cytogenetic assays will be applicable as biomarkers of toxic effect for food chemicals where the exposure levels are much lower.

3.2.2 p53 Gene Mutations. Alterations in the p53 tumour suppressor protein are a common feature of many cancers and about half of all cancer cases involve mutation of the gene.[46] The p53 protein, but not mutant forms, appears to control the cell cycle checkpoint responsible for maintaining the integrity of the genome and induces cell cycle arrest in response to DNA damage.[47]

Maestro and co-workers[48] have recently investigated the incidence of p53 protein overexpression and the frequency of p53 gene mutations in squamous cell carcinomas from 58 smokers. In some 60% of cases overexpression of nuclear p53 protein was detected as revealed by immunohistochemical staining with monoclonal antibodies. To ascertain the reasons for this overexpression gene sequencing studies were undertaken on a number of samples. This revealed mutations in the p53 gene of all 6 samples showing overexpression that were analysed but in none of the 11 samples analysed where overexpression was absent. The high incidence of mutations in the p53 gene suggests that mutationally-derived inactivation of the tumour suppressor protein is likely to be a key factor in the progression of these cancers.

The case of Hubert Humphrey, vice-president of the USA from 1963 to 1969 illustrates the potential role of p53 mutations as a biomarker of toxic effect.[49] In 1967 Humphrey noted blood in his urine and consulted his physician. Examination of the bladder tissue proved inconclusive and a "wait and see" policy was adopted. Borderline malignancy was diagnosed in 1973 and death from bladder cancer occurred in 1978. Following storage of the bladder tissue samples they were recently analysed for mutations in the p53 gene utilising the polymerase chain reaction amplification technique, which was not available at the time that Humphrey was alive. This revealed the presence of telltale mutations in the primary bladder carcinoma surgically removed in 1976. Intriguingly one of these p53 mutations was also detected in the original biopsy sample taken in 1967, i.e. 11 years before his death.

Mutations in the p53 gene in tumours are not randomly distributed across the gene and the majority are clustered between codons 130 to 290 with "hot spots" occurring at specific sites.[47] In addition the profile of nucleotide transversions is not constant but varies depending upon the organ. Utilising data from a number of studies Puisieux *et al*[50] have shown that G:C to T:A transversions predominate in lung (46%) and liver (92%) tumours whereas G:C to A:T (76%) are the major transversions in colon cancers. The reasons underlying these tissue specific mutations are not clear. One hypothesis is that they may in part reflect localised exposure to different carcinogens. In this context it is interesting to note[50] that both benzo(*a*)pyrene (BaP) and AFB1 preferentially adducted to the guanine bases of the p53 gene *in vitro*. Treatment with epoxides of BaP and AFB1 resulted in 87% and 62%, respectively, of guanine bases being adducted when the p53 gene was exposed to these carcinogens. In contrast interaction with other bases was significantly less, e.g. only 11% of adenine bases were adducted on treatment with BaP and 3% on exposure to AFB1. However whilst BaP and AFB1 both preferentially form adducts with guanine bases they demonstrate differential effects when the position of the nucleotide on the p53 gene is considered. Thus whereas AFB1 causes a G to T mutation at the third base of codon 249 BaP does not. This is of interest in so much as the 249 mutation in the p53 gene, which is rarely seen in liver cancers in Western countries, was detected in 55% of liver cancers from Quidong, China, an area of high exposure to AFB1.[51] In a related study[52] in Quidong lower, but detectable, levels of p53 mutations have been found in non malignant liver tissue from patients with hepatocellular carcinomas. The presence of these mutations in adjacent, non malignant, tissue suggests that p53 mutations may occur early in the carcinogenic process. Conversely p53 mutations detected in liver in other studies are thought to occur late in tumorigenesis. More work is required to assess the influence of the sensitivity of the assay methods employed before firm conclusions can be drawn.

4 CONCLUSIONS

Whilst the application of biomarker-based approaches to risk assessment of food chemicals is a relatively new area significant advances have been made with respect to biomarkers of exposure. Less progress is evident for biomarkers of toxic effect. This is in large measure due to the incomplete understanding of the molecular factors underlying the progression of many disease states. As a consequence it seems unlikely that biomarkers will replace the existing, animal-based, approaches to risk assessment in the near future. However biomarkers have considerably more potential, in terms of more accurately reflecting the actual risks that food chemicals pose to humans, and this potential will no doubt be increasingly realised as a result of future R&D in this area.

5 REFERENCES

1   ECETOC, DNA and protein adducts: evaluation of their use in exposure monitoring and risk assessment, European Chemical Industry Ecology and Toxicology Centre, Brussels, 1989, No. 13.

2.   National Research Council Committee on Biological Markers, *Environ. Health Perspect.*, **74**, 3.

3.   Ministry of Agriculture, Fisheries and Food, *The British Diet: Finding the Facts. Food Surveillance Paper No. 23,* H.M.S.O., London, 1988.

4.   M. J. Heard, A. C. Chamberlain and J. C. Sherlock, *Sci. Total Environ.*, 1983, **30**, 245.

5.   J. D. Groopman, J. Zhu, P. R. Donahue, A. Pikul, L.-S. Zhang, J.-S. Chen and G. N. Wogan, *Cancer Res.*, 1992, **52**, 45.

6.   M. E. Peattie, D. H. Buss, D. G. Lindsay and G. A. Smart, *Food Chem. Toxicol.*, 1983, **21**, 503.

7.   K. Ungar, S. Minors and M. Balls, 'A feasibility study on developments in biomarkers of potential utility in risk management and risk assessment of food chemicals', Fund for the Replacement of Animals in Medical Experiments (FRAME), Nottingham, 1992.

8.   J Lewalter, U Korallus and H Weidermann, *Int. Arch. Environ. Health*, 1985, **55**, 33.

9.   A. M. Lynch, S. Murray, A. R. Boobis, D. S. Davies and N. J. Gooderham, *Carcinogenesis*, 1991, **12**, 1067.

10.   R. M. Santella, *Mutat. Res.*, 1988, **205**, 271.

11.   M. V. Reddy and K.Randerath, *Carcinogenesis*, 1986, **7**, 1543.

12.   N. J. Gorelick and G. N. Wogan, *Carcinogenesis*, 1989, **10**, 1567.

13.   B. P. Dunn, *Environ. Health Perspect.*, 1991, **90**, 111.

14.   Ministry of Agriculture, Fisheries and Food, *Nitrate, Nitrite and N-Nitroso Compounds: Second Report. Food Surveillance Paper No. 32*, H.M.S.O., London, 1992.

15.   P. Correa, W. Haenszel, C. Cuello, D. Zvacalal, E. Fontham, G. Zarama, S. Tannenbaum, T. Collazos and B. Ruiz, *Cancer Res.*, 1990, **50**, 4731.

16.   P. J. Packer and S. A. Leach, 'Nitrate and Nitrites in Food and Water'; M. J. Hill (ed.), Ellis Horwood, London, 1991, p. 131.

17.   P. J. Packer, C. P. J. Caygill, M. J. Hill and S. A. Leach, *Aqua*, 1995, in press.

18.   National Toxicology Program, Carcinogenesis bioassay of di(2-ethylhexyl) adipate (CAS No. 103-23-1) in F344 rats and B6C3FI mice (feed study), Research Triangle Park, N.C., U.S.A., *Technical Report Series No. 212*, 1980.

19. Ministry of Agriculture, Fisheries and Food, *Plasticisers: Continuing Surveillance. Food Surveillance Paper No. 30*, H.M.S.O., London, 1990.

20. N. J. Loftus, W. J. D. Laird, G. T. Steel, M. F. Wilks and B. H. Woollen, *Food Chem. Toxicol.*, 1993, **31**, 609.

21. N. J. Loftus, B. H. Wollen, G. T. Steel, M. F. Wilks and L. Castle, *Food Chem. Toxicol.*, 1994, **32**, 1.

22. G. N. Wogan, 'Aflatoxin Scientific Background, Control and Implications. Food Science and Technology. A Series of Monographs', L. A. Goldblatt, (ed.), Academic Press, London, 1969, p. 151.

23. F. C. Jelinek, A. E. Pohland and G. E. Wood, *J. A.O.A.C.*, 1989, **72**, 223.

24. R. C. Garner, R. Ryder and R. Montesano, R., *Cancer Res.*, 1985, **45**, 922.

25. J. Gilbert, *Food Addit. Contam.*, 1993, **10**, 37.

26. J.-Q. Zhu, L.-S. Zhang, X, Hu, Y. Xiao, J.-S. Chen, Y.-C. Xu, J. Fremy and F. S. Chu, *Cancer Res.*, 1987, **47**, 1848.

27. A. Zarba, C. P. Wild, A. J. Hall, R. Montesano and J. D. Groopman, *Carcinogenesis*, 1992, **13**, 891.

28. S. Tsuboi, T. Nakagawa, M. Tomita, T. Seo, H. Ono, K. Kawamura and N. Iwamura, *Cancer Res.*, 1984, **44**, 1231.

29. J. M. Essigmann, R. G. Croy, A. M. Nadzan, W. F. Busby Jr., V. N. Reinhold, G. Buchi and G. N. Wogan, *Proc. Natl. Acad. Sc. U.S.A.*, 1977, **74**, 1870.

30. J. D. Groopman, J. Hasler, L. J. Trudel, A. Pikul, P. R. Donahue and G. N. Wogan, *Cancer Res.*, 1992, **52**, 267.

31. G. Sabbioni, P. Skipper, G. Buchi and S. R. Tannenbaum, *Carcinogenesis*, 1987, **8**, 819.

32. C. P. Wild, R. G. Garner, R. Montesano and F. Tursi, *Carcinogenesis*, 1986, **7**, 853.

33. L.-S. Gan, P. L. Skipper, X. Peng, J. D. Groopman, J.-S. Chen, G. N. Wogan and S. R. Tannenbaum, *Carcinogenesis*, 1989, **9**, 1323.

34. J. D. Groopman, A. Hall, H. Whittle, G. Hudson, G N. Wogan, R. Montesano and C. P. Wild, *Cancer Epidemiol. Biomarkers Prev.*, 1992, **1**, 221.

35. C. P. Wild, G. Hudson, G. Sabbioni, G. N. Wogan, H. Whittle, R. Montesano and J. D. Groopman, *Cancer Epidemiol. Biomarkers Prev.*, 1992, **1**, 229.

36. B. S. Hulka, *Arch. Environ. Health*, 1988, **42**, 83.

37. V. J. McKelvey-Martin, M. H. L. Green, P. Schezer, B. L. Pool-Zobel, M. P. DeMeo and A. Collins, *Mutat. Res.*, 1993, **288**, 47.

38. R. Waters, 'Mutagenicity Testing - A Practical Approach', S. Venitt and J. M. Parry, (eds.), I.R.L. Press, London, 1984, p. 99.

39. P. W. Brandt-Rauf, *Int. Arch. Occup. Environ. Health*, 1991, **63**, 1.

40. B. Bressac, M. Kew, J. Wands and M. Ozturk, *Nature (London)*, 1991, **350**, 429.

41. I. Alder, 'Mutagenicity Testing - A Practical Approach', S. Venitt and J. M. Parry, (eds.), I.R.L. Press, London, 1984, p. 275.

42. M. L. Cleary, *Cell*, 1991, **66**, 619.

43. E. J. Stanbridge, *Cancer Surv.*, 1992, **2**, 5.

44. G. Tong, N. J. Van Sittert and A. T. Natarajan, *Mutat. Res.*, 1988, **204**, 451.

45. L. Hagmar, A. Brogger, I.-L. Hansteen, B. Hogstedt, L. Knudsen, B. Lambert, K. Linnainmaa, F. Mitelman, I. Nordsenson, C. Reuterwall, S. Salomaa, S. Skerfving and M. Sorsa, *Cancer Res.*, 1994, **54**, 2919.

46. M. Hollstein, D. Sidransky, B. Vogelstein and C. C. Harris, *Science*, 1994, **253**, 49.

47. A. J. Levine, J. Momand and C. A. Finlay, *Nature (London)*, 1991, **351**, 453.
48. R. Maestro, R. Dolcetti, D. Gasparotto, C. Doglioni, S. Pelucchi, L. Barzan, E. Grandi and M. Boiocchi, *Oncogene*, 1992, **7**, 1159.
49. B. J. Culliton, *Nature (London)*, 1994, **369**, 13.
50. B. Puisieux, S. Lim, J. Groopman and M. Ozturk, *Cancer Res.*, 1991, **51**, 6185.
51. I. C. Hsu, R. A. Metcalf, T. Sun, J. A. Welsh, N. J. Wang and C. C. Harris, *Nature (London)*, 1991, **350**, 427.
52. F. Aguilar, C. C. Harris, T. Sun, M. Hollstein and P. Cerutti, *Science*, 1994, **264**, 1317.

Biomarkers Used to Validate Dietary Assessments in Human Population Studies

Sheila A. Bingham

MRC DUNN CLINICAL NUTRITION CENTRE, HILLS ROAD, CAMBRIDGE CB2 2DH, UK

1 INTRODUCTION

Methods used to assess food intake in free living individuals include food diaries, weighed food records and questionnaires. Food frequency questionnaires (FFQ) are particularly used in nutritional epidemiology, since they are designed to assess the habitual food and nutrient intake of large numbers of people relatively rapidly. 24 h recalls are also commonly used but their limitations for individual assessments due to attenuation from daily variation in nutrient intake are well known [1]. However, the validity of measurements of dietary intake in free living individuals is difficult to assess because all methods rely on information given by the subjects themselves, which may not be correct. In an attempt to determine objective measures of validating dietary assessments, biological specimens that closely reflect dietary intake, but which do not rely on reports of food consumption are now in use [1].

The doubly labelled water technique can be used as a non invasive measure of energy expenditure, and, with no gains or losses in weight, expenditure should equal energy intake. Studies using this method are however limited by the expense of the technique, and comparisons may be limited to the energy intake to basal metabolic rate (BMR) ratio, calculating BMR from body weight, and with assumptions about energy expenditure [2].

To assess the utility of the 24 h urine nitrogen technique to validate individual measures of dietary intake, and to develop a marker to verify the completeness of 24 h urine collections, initial work was conducted in a metabolic suite [3,4]. This showed that a minimum of 16 days of weighed records and 8 24 h urine collections were required to assess usual dietary intake, and subsequent validation protocols therefore specified 4 days of weighed food records and two 24 h urine collections, to be repeated at 3 month intervals over a year.

The UK arm of the European Prospective Investigation of Cancer (EPIC), is a long term study in which habitual dietary patterns of which fifty thousand individuals aged 55 to 75 years is being assessed. The members of the cohort are then to be followed for over 10 years, when the dietary habits of the participants who succumb to cancer (or other conditions such as heart disease) can be compared with those who do not. Blood samples (serum, plasma, white and red cells) are also being obtained for storage in liquid nitrogen until end point analysis.

To determine the accuracy of methods for dietary assessment suitable for such large numbers of participants an initial study was conducted in which the results from methods usually used to assess diet in prospective studies were compared with those from the validation protocol. The aim was to find a method that was able to assess usual diet. The chosen method had also to cover intake of all items of diet since most have been linked to initiation, promotion, and prevention of cancer, and be flexible enough to cope with changes in hypotheses as the cohort progressed over time.

2 METHODS

Each individual was studied at home on four occasions (seasons) over the course of one year, and at each season was asked to complete 4 days of weighed food records. The volunteers were also asked to provide two 24 h urine collections on each occasion, so that over the year each individual provided 16 days of weighed dietary records, and eight 24 h urine collections. Seven dietary methods were investigated over the course of one year, two different FFQs, two variants of the 24 h recall, and three types of food diary. Details of the methods tested have been described elsewhere [5,6].

2.1 Weighed records

A set of dietary balances, called PETRA (Portable Electronic Tape Recorded Automatic) scales were developed to minimise the burden on volunteers when asked to weigh and record their food for prolonged periods of time (Cherlyn Electronics Limited, Cambridge). These are accurate to $\pm$ 1 g and automatically record verbal descriptions and weights of food on a dual track cassette, thus avoiding the necessity for subjects to keep written records. PETRA is of additional benefit in dietary validation studies because weights of foods are recorded on tape and not disclosed to the participant [1]. Each 4-day period included different days chosen to ensure that all days of the week were studied, and that there was an appropriate ratio of weekend to weekdays included during the year. The weights and verbal descriptions of food consumed were transposed by hand from the tape cassettes using the PETRA master console, and hand-coded for computer calculation of nutrient intake using food tables [5,6].

2.2 Urine collections, anthropometry and blood sampling

Subjects asked to make the first 24 h urine collection on the third or fourth day of weighing their food, and the second within one week of completing the 4 day records. For each collection, they were given a canvas bag containing two 2 l containers, each with 2 g boric acid as preservative, a jug, funnel, safety pin to pin to their underclothes as a reminder, and a set of instructions. They were asked to discard the first specimen of the day, for example at 7 am, and from then on to collect all specimens for 24 h up to and including 7 am the next day. They were given three 80 mg tablets of p- amino benzoic acid (PABAcheck, Laboratories for Applied Biology, London) to take with meals as a marker for completeness of 24 h urine collections ([4,7]). Overnight fasting blood samples were also obtained, together with body weight and fasting breath samples at each occasion. Analytical methods were as published[7].

2.3 FFQs, 24 h recalls and food diaries

Two FFQs were tested, each with 130 food items, each asking subjects to recall "average use last year" and each using the same frequency categories of use. The first (Cambridge) questionnaire was sent to subjects by post as part of recruitment before they participated in the dietary studies. The list of foods included in each category of this questionnaire was chosen from foods most commonly eaten in a previous study of a Cambridge population. No units or portion weights were specified and average portion weights were used from the first 4 day weighed records kept by 80 of the participants. The portion size for milk was 59 g. The second (Oxford) questionnaire was given to subjects to complete immediately before they started to weigh their food on day 0 in season 3 and was based on an American FFQ [5]. 'Medium serving' or units were specified (pints, slices, teaspoons, etc.) and the portion size assigned to milks was 567 g (1 pint). Full details are published elsewhere [5].

Two types of 24 h recall were tested. The first was unstructured and consisted of a blank sheet of paper and written example attached to the back of the Cambridge FFQ, which was given to the subjects during recruitment before the start of the study. Published portion weights were used to calculate nutrient intakes.The second was a structured 24 h recall given as a booklet for subjects to fill in just before starting the weighed records in season 2 and entailed detailed enquiries as to the amount of food eaten and portion sizes using a set of photographs [5].

Three 7-day estimated diet records (diet diaries) were tested. All were left with the subjects to fill in within 14 days of completing the 4 day weighed records, with stamped addressed envelopes for postal return. In season 1, subjects received a 7 day diary checklist. This comprised one page of instructions, one of an example, and seven pages (one for each day) of the checklist. The checklist was a printed list of 160 foods, largely that of the Cambridge FFQ, on which subjects were asked to check off which foods they had eaten, counting half for a small portion, and two for a large portion at the end of each day. In season 4, a modified version of the checklist designed to obtain individual estimates of portion size using photographs was used. In season 3, 80 subjects completed an open ended (unstructured) estimated diet record or 7-day food diary booklet. Subjects gave written detailed descriptions of food consumed at the time of eating using 15 sets of photographs printed in the booklet to establish portion size. Subjects were allowed to state portion sizes in other measures if they so wished. No part of this diary was precoded but a computer programme, DIDO (Data In, Diet Out) was developed specifically for the coding of these booklets for nutrient analysis [5].

3 RESULTS

3.1 Comparisons between weighed records, the FFQs, 24 h recalls and food diaries

Both FFQ yielded significantly greater intakes of fibre non -starch polysaccharides, potassium, carotene and vitamin C, due to a greater reported frequency of consumption of vegetables, compared with the weighed records. Intakes of vegetables were reported as 406 and 386 g per day by the FFQs, compared with 272 g from weighed records. As a result of this difference of 120 g per day of vegetables, carotene levels estimated by FFQs were as high as 5.1 and 4.7 mg per day, compared with 3.4 mg per day from weighed records. Average intakes of milk were 490 g from the Oxford FFQ, but 175 g per day from the Cambridge FFQ, compared with 296 g per day from the weighed records. Hence average intakes of calcium, sugars, fat and protein were significantly greater when assessed by the Oxford questionnaire, but significantly lower by the Cambridge questionnaire.This finding, that FFQ results yield higher average daily intakes, particularly of vegetables, than other methods has been noted in other comparative studies [8]. There were few differences in mean intakes assessed by the unstructured 24 h recall and checklist methods. There were no significant differences in mean intakes of nutrients or foods assessed from the 7 day diet records compared with weighed records [5].

Spearman rank correlation coefficients between the individual values obtained from each method and the averages of all 16 day weighed records were highest for alcohol (0.54 to 0.90), and lowest for protein, iron and carotene (0.13 to 0.67). The Cambridge FFQ performed least well on this analysis, with coefficients ranging from 0.13 to 0.41 excluding alcohol, compared with 0.39 and 0.57 in the Oxford questionnaire. Correlation coefficients were of the same order for both 24 h recall methods, and ranged from 0.21 to 0.60 in the simple method and from 0.14 to 0.65 in the structured method with pictures to assess portion size. Correlation coefficients were highest for all three record methods, and tended to be of a greater magnitude between the 16 days of weighed records and the 7 day open ended food diary [5].

3.2 Validity of weighed records and 'under reporting'

There were highly significant differences in total nitrogen, potassium, sodium, and volume between the samples designated complete and incomplete by the PABA technique[6]. To assess the validity of the dietary assessment methods therefore, only those 24 h urine collections that were complete were used in these analyses. Average N intake from the 16 day records was 11.2 ± 2.3 g per day, and that from N excretion in the complete 24 h urine 9.84 ± 1.78 g per day, so that the average ratio of urine N to dietary N was 0.91 ± 0.09. This was rather greater than the ratio of 0.81 ± 0.05 expected if the average results from all individuals were valid [6].

To determine which, if any, of the individual results were valid, the ratio of urine to dietary N was sorted and data was examined as quintiles of the distribution of the urine to dietary nitrogen ratio. Means of this ratio ranged from 0.76 in the lower quintile of the distribution to 1.13 in the upper quintile. Examination of correlations between urine and dietary N, ratios of energy intake to BMR, correlations of the ratio of energy intake to BMR with the urinary to dietary N ratio, body mass index (BMI), and body weight, indicated that mean values from the 20% individuals assigned to the top quintile were different from data from the 80% individuals assigned to the other four quintiles [6].

All data were therefore considered separately for individuals in the top quintile and for individuals in the other four quintiles of the distribution in urine to dietary N ratio. Not only were individuals in the top quintile heavier, with a lower energy intake to BMR ratio than the others, but their intakes of energy and all energy yielding nutrients calculated from weighed records were significantly lower than those from individuals in the other quintiles. On average there was an 18 g difference in reported fat consumption, and a 27 g difference in reported sugars consumption between the average values reported in the top and the other four quintiles according to the urine to dietary N ratio. Mean consumption of cakes, breakfast cereals, milk, eggs, fats, and sugars was also significantly lower in those individuals classified in the top quintile of the distribution. Under reporting did not seem to be limited to weighed records, since reported intakes of energy and energy yielding nutrients by these individuals was as likely when they used another dietary assessment method, the Oxford FFQ [6].

However, there was no difference in reported consumption of meat, fruits, vegetables and potatoes between these under reporters and the other 80% of the population who gave valid records. Correspondingly, there was no difference in plasma vitamin C between the two groups (62 (se)2 , 61 (se) 5) umol/l) although plasma carotenoids were significantly lower in the top quintiles, probably due to the fat solubility of carotenoids and the greater body fat content of individuals in the top quintile [8].

3.3 Comparisons between assessments from different methods and biological indices of diet

Correlations between dietary variables from the 16 day weighed record averages with mean values of analytes in urine, blood and breath were calculated. As well as the expected correlation with 24 h urine N and dietary N from the weighed records ($r = 0.69$), there was also a high correlation between 24 h urine potassium excretion and dietary potassium intake, $r = 0.76$. Despite the fact that estimates of carotene, mainly β carotene, only were available in the data base to calculate carotene intake, there were significant correlations between diet estimates of carotene and plasma concentrations of carotenoids ranged from 0.20 for plasma β cryptoxanthin, to 0.62 for plasma α carotene. The correlation between plasma vitamin C and dietary vitamin C with values from supplements included, was 0.86 (Spearman rank coefficient), but that between intake from weighed records alone (i.e. excluding supplements) and plasma 0.32. Plasma vitamin C was not considered further, due to the effect of added vitamin C supplements.

The validity of dietary methods was then considered by comparisons with these biological markers, 24 h urine N, 24 h urine K, and plasma carotenoids. The table shows correlation coefficients, both in the group as a whole and in the 80% individuals who had submitted valid records as judged by the urine to dietary nitrogen ratio. Correlation coefficients were highest between urine N and dietary N weighed records (0.69 all values, 0.87 valid records) and correlations with the 7 day diary records were also high (0.65 all values, 0.70 valid records). Correlations with checklist methods were in the order of 0.4, and correlations between urine and dietary N from FFQs and 24 h recalls were much lower, 0.10 and 0.24. There did not appear to be differences between groups of individuals classified into the top versus other four quintiles, apart from a negative correlation (-0.31) between urine and dietary N in the top quintile of 24 h recall data and a positive correlation (0.50) in the top quintile of the Oxford FFQ.

Correlations between potassium intake estimated from the weighed records, food diaries, FFQ and 24 h recalls showed the same general pattern as those found with 24 h urine nitrogen. Correlations were highest (0.74 - 0.82) for the weighed records, next best for the 7 day diary (0.66), followed by the checklists (0.44-0.49), 24 h recalls (0.51-0.53), and much lower (0.25 - 0.28) for the FFQs.

Correlations between carotene intake estimated from weighed records and plasma carotenoids were also much higher between than those with values derived from the FFQs and 24 h recalls. The correlation between plasma β carotene and dietary carotene from weighed records was 0.46 for example, whereas it was only 0.15 and 0.04 using values for dietary carotene from the FFQs. There was little change in the correlations when considered by top versus lower four quintiles. Plasma carotenoids were not analysed in the individuals who completed the 7 day diary, so a direct comparison with the 7 day diary is not possible. However, correlations with carotene intake assessed from a similar method, the structured 7 day food checklist are shown in the Table. In general, correlations between dietary carotene intake using this method and plasma β carotene were higher than those obtained with intakes derived from the 24 h recalls and FFQs. The correlations between plasma β carotene and dietary carotene from this method for example was 0.25, 0.51 and 0.30, lower than those with weighed records, but higher than those found with the FFQs and 24 h recalls.

4 DISCUSSION

In analytic epidemiological studies, the ability of methods to place individuals in the correct place in the distribution of habitual food and nutrient intake is the most important requirement for the estimation of relative risk and trend analysis. FFQs are the most commonly used dietary assessment technique used in large scale prospective studies. It was therefore surprising to find that the FFQ was not appreciably better than other methods at placing individuals in the distribution of habitual intake for the majority of nutrients, either when compared with weighed records of food intake or biomarkers, as judged by correlation analysis. When compared with weighed records, high correlations were only obtained for alcohol consumption in the FFQ, but this is a common finding due to the fact that a substantial proportion of the individual values obtained are zero, leading to a wide range in intake, and that alcohol is generally consumed in standard units [5].

Despite often expressed doubts about the accuracy of the weighed intake technique, results from it were remarkably well correlated with two biological markers of food intake, the 24 h urine N and K (Table). In fact, it easily outperformed the FFQ and single 24 h recall methods in correlation analyses when compared with 24 h urine N and K, and plasma carotenoids. This was true in the whole group, the 80% who gave valid records,

Table Pearson correlations between dietary intake assessed from dietary methods with biomarkers of intake

	1 - 4th Quintile	Top Quintile	All values
Correlations between intake of Nitrogen (N) from methods versus 24 h urine N			
N from weighed records	0.87	0.78	0.69
N from 7 day diary	0.70	0.60	0.65
N from checklist	0.45	0.32	0.38
N from checklist with portions	0.43	0.26	0.38
N from Oxford questionnaire	0.27	0.50	0.24
N from Cambridge questionnaire	0.15	0.19	0.15
N from unstructured 24 h recall	0.26	-0.31	0.10
N from structured 24 h recall	0.25	0.01	0.21
Correlations between intake of Potassium (K) from methods versus 24 h urine K			
K from weighed records	0.82	0.74	0.76
K from 7 day diary	0.70	0.57	0.66
K from checklist	0.49	0.49	0.49
K from checklist with portions	0.45	0.45	0.44
K from Oxford questionnaire	0.28	0.28	0.25
K from Cambridge questionnaire	0.21	0.34	0.24
K from unstructured 24 h recall	0.52	0.53	0.51
K from structured 24 h recall	0.49	0.22	0.41
Correlations between intake of carotene from methods versus plasma β carotene			
Carotene from weighed records	0.45	0.46	0.46
Carotene from checklist	0.25	0.51	0.30
Carotene from checklist with portions	0.28	0.23	0.27
Carotene from Oxford questionnaire	0.15	-0.02	0.15
Carotene from Cambridge questionnaire	0.07	-0.02	0.04
Carotene from unstructured 24 h recall	0.06	0.10	0.09
Carotene from structured 24 h recall	-0.01	-0.18	0.00

and the 20% of overweight individuals who did not. This finding of a better relationship between 24 h urine N and records, rather than FFQ has been observed elsewhere. Correlations between 24 h urine nitrogen from a single collection and protein intake assessed from a 4 day record were 0.47, and that between the 24 h urine nitrogen and a FFQ, 0.19 [9].

The errors associated with known fluctuations in the frequency of consumption of micro nutrients such as the carotenes are reduced by repeated observations with 24 h recall or record methods, but the FFQ is designed to assess usual intake without the need for repeated assessments. Direct comparisons with the estimated menu record (7 day diary) were not possible, because plasma carotenoids were not analysed in this group. However, data from a similar method, the checklist, were available and gave higher results than those obtained from the FFQ, for example 0.15 for β carotene assessed by FFQ, and 0.28 for β carotene assessed from the record (Table). Other studies have also suggested that carotene intakes assessed from food records correlate to a higher degree with plasma carotene than those from FFQ[10]. So far therefore, the low correlations between results obtained from the FFQ and serum carotenes suggest that the errors associated with attempting to assess frequency of usual intake from questionnaires are greater than those from random day to day variations associated with daily records or 24 h recalls kept over comparatively short periods of time. Hence, the errors associated with known fluctuations in the frequency of consumption of micro nutrients cannot be entirely overcome by using a FFQ designed to assess usual intake without the need for repeated assessments.

Overall, the unstructured 7 day diary (estimated record) had the highest correlation coefficients and was able to classify a greater proportion of individual values into the correct place of the distribution when compared with weighed records. No biases in mean intakes of either foods or nutrients were found, and comparisons between results obtained from it and biological markers of intake were almost as good as those obtained from 16 days weighed records. For these reasons, and because of the flexibility of a food diary in investigating any further hypotheses, it is to be used as the main method in the UK arm of the EPIC study, with repeat investigations as the cohort progresses over time.

6 REFERENCES

1. Bingham S, The dietary assessment of individuals; Methods, accuracy, new techniques and recommendations, Nutrition Abstracts and Reviews ,1987, **5 7**, 705-742
2. Goldberg, G.R., Black, A.E., Jebb, S.A., Cole, T.J., Murgatroyd., P.R., Coward, W.A., Prentice, A.M, Critical evaluation of energy intake data : 1 Derivation of cut off limits to identify underrecording, Eur J Clin Nutr., 1991, 45,569-581
3. Bingham, S., Cummings, J.H., The use of 4 amino benzoic acid as a marker to validate the completeness of 24 h urine collections in man , Clinical Science, 1983, 1983,64 629-635
4. Bingham, S., Cummings, J.H., Urine Nitrogen as an independent validatory measure of dietary intake Am J Clin Nutr , 1985, 42,1276-1289
5. Bingham, S.A., Gill, C., Welch, A., Day, K., Cassidy, A., Khaw, K.T., Sneyd, M.J., Key, T.J.A., Roe, L., Day, N.E., Comparison of dietary assessment methods in nutritional epidemiology B J Nutr., 1994, 72,619-643
6. Bingham, S.A.,Cassidy, A., Cole T., Welch A., Runswick, S., Black A.E., Thurnham, D., Bates, C.E., Cassidy, A., Khaw, K.T. & Day, N.E., Validation of weighed records and other methods of dietary assessment using the 24 h urine technique and other biological markers B.J.Nutr, 1995, 73,531-550
7. Bingham, S., Williams, D.R.R., Cole, T.J., Price, C.P. & Cummings, J.H. Reference values for analytes of 24 h urines known to be complete. Ann. Clin. Biochem. 1988, 25, 610-619.
8. Bingham, S. A., Gill, C., Welch, A, Cassidy, A, Runswick, S.A., Sneyd M.J., Thurnham, D., Key T.J.A., Roe, L, Khaw, K. T., Day, N. E., Validation of dietary assessment methods in the UK arm of EPIC using weighed records, and 24 h urinary nitrogen and potassium and serum carotenoids as biomarkers.Int J Epid (in press)
9. Rothenberg, E., Validation of the FFQ with the 4 day record method and analysis of 24 h urinary nitrogen Eur J Clin Nutr., 1994, 48 725-735
10. Yong, L.C., Forman, M., Beecher, G.R., Graubard, B.I., Campbell, W.S.,Reichman, M.E., Taylor, P.R., Lanza, E., Holden, J.M., Judd, J.T., Relationship between dietary intake and plasma concentration of carotenoids in premenopausal women, Am J Clin Nutr , 1994, 60,223-30

Acknowledgements
Support for this study was obtained from the Medical Research Council, the Cancer Research Campaign, the Ministry of Agriculture, Fisheries and Food, The Imperial Cancer Research Fund, and the Vegetarian Society.

Biological Monitoring of Trace Elements to Indicate Intake and Uptake from Foods and Beverages

H. T. Delves

DEPARTMENT OF CLINICAL BIOCHEMISTRY, THE UNIVERSITY OF SOUTHAMPTON, SOUTHAMPTON SO16 6YD, UK

1 INTRODUCTION

Biological monitoring is a well established technique for assessing increased intakes and uptakes of toxic elements following either occupational or environmental exposure. Relative intakes and uptakes of both essential, and non-essential, trace elements from foods may also be assessed by measuring elemental concentrations in body fluids. Although, in the case of essential trace elements the changes in concentration in response to hormonal influences and metabolic disorders will need to be considered. Examples given here are the use of elemental concentrations in body fluids as biomarkers of excessive intakes from foods: these include manganese, an essential micronutrient and the non-essential elements, aluminium, mercury and lead.

2 MANGANESE

This nutritionally essential trace element is an integral component of many enzymes, including pyruvate carboxylase and galactosyl transferase. Manganese deficiency is associated with abnormalities of mucopolysaccharide synthesis and with defective formation of connective tissues, whereas manganese toxicity is associated with neurological disorders involving degeneration of the basal ganglia with predominant gliosis of the globus pallidus and Parkinsonism.[2,3] Excessive industrial exposure via inhalation of oxide dusts can lead to toxicity states which are manifest by Parkinson-like tremors. Recently manganese toxicity, associated with neurological disturbances, has been associated with excessive oral intakes from contaminated foods[4] and excessively high concentrations of manganese in intravenous feeding solutions.[5]

Contamination of soil from a surface manganese-ore deposit in Groote Eylandt, Northern Territories of Australia, increased the concentrations of manganese in locally grown fruit and vegetables, by up to 700 fold (Table 1).[4] The local aboriginal population who consumed these foods had considerably elevated concentrations of manganese in their blood: median 450nmol l^{-1}, range 175 to 900nmol l^{-1} compared with a reference population of non-exposed city dwellers whose levels were: mean 215nmol l^{-1}, range 85-350nmol l^{-1}. A subset of about 2% of the aboriginal population with blood manganese levels > 500nmol l^{-1} had neurological disorders.

Overt manganese toxicity was recently reported in a 62 year old man who had received total parenteral nutrition for 31 months.[5] He was admitted to hospital with an ataxia which had progressively worsened over the latter 16 months. His clinical symptoms included signs of Parkinsonism with: dysarthria, mild rigidity, hypokinesia with a masked face, a halting gait and severe postural reflexes. His whole-blood manganese had been within the range 546-1018 nmol l^{-1}, well in excess of a reference range 73-364 nmol l^{-1}. Other laboratory findings were normal except for a slight renal dysfunction. A 1.5T magnetic resonance imaging (MRI) scan showed enhanced signals in the basal ganglia and brain-stem. Deposition of manganese in brain tissue produces enhanced MRI signals via its strong paramagnetism which shortens spin-lattice relaxation times. On cessation of manganese administration his Parkinsonism gradually regressed, the MRI signals declined and his whole-blood manganese fell to 91nmol l^{-1}.

The similarity in the whole-blood manganese concentrations associated with neurological disorders in both the aboriginal population and in the subject who had received TPN with high manganese intake is striking; in both studies toxicity was associated with concentrations above 500nmol l^{-1}, and in both studies the reference populations were below 360nmol l^{-1}. It has been known for many years that patients receiving total parenteral nutrition (TPN) can have abnormally high concentrations of manganese in serum[6] and in whole-blood.[7] However it is only recently that the toxicity associated with high manganese concentrations has been appreciated. The frequency of abnormally high blood manganese concentrations in adults and in children receiving TPN in two UK hospitals are shown in Figures 1,2. About 75% of adults from Southampton General Hospital and about 95% of children from the Hospital for Sick Children, Great Ormond Street Children's Hospital had blood manganese concentrations above the upper limit of normal of 210nmol l^{-1}. Potentially toxic concentrations, >360nmol l^{-1} were seen in 38% of 161 adults at Southampton and in 61% of 57 children at Great Ormond Street. Of these 3% of adults and 19% of children could be considered at risk of neurological damage. The TPN solutions provided manganese intakes greatly in excess of daily requirements: up to 10x more for adults and up to 50x more for children. These results were conveyed, by the author, to the Department of Health in February 1994 and immediate action was taken to minimise the exposure of TPN patients to high manganese intake from TPN solutions.[8]

The use of manganese determinations was recently questioned by Halls[9] who particularly queried the choice of a reliable index of manganese status *e.g.* whole-blood, or plasma, or hair, or nails. Although the concentrations of manganese in both plasma and whole-blood increase with increasing intake the whole-blood concentration is 5 to 10x greater than plasma and is therefore more likely to be a more sensitive index. Milne et al. suggested whole-blood or lymphocytes as indicators of manganese status.[10] It is clear from the data in Figures 1,2 and from the earlier discussions that measurement of whole-blood manganese provides a valid assessment of the uptake of manganese from foods. Whole-blood manganese is a relatively simple determination by electrothermal atomisation and atomic absorption spectrometry (ETA-AAS) and requires only 50-100 μl sample volumes. The use of good quality control can ensure that the accuracy and precision of these assays is within the range 3 to 5%.

Table 1   **MANGANESE IN FRUIT/VEGETABLES GROWN ON CONTAMINATED SOILS[5]**

Fruit/Vegetable	Mn, μg g^{-1}
Banana	31
Paspalum	240
Yam	720
World-wide Reference Data	0.2 - 7.7

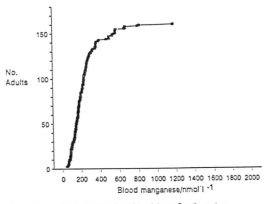

Figure 1   *Blood Mn data for 161 adults at Southampton*

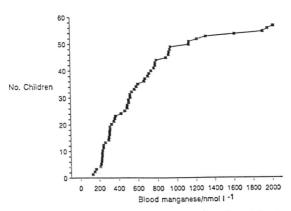

Figure 2   *Blood Mn data for 57 children at Great Ormond Street*

3   ALUMINIUM

The concentrations of aluminium in foods and beverages range from less than $1\mu g$ kg^{-1} to greater than 1g kg^{-1} depending upon the type of food, the degree of contamination during manufacture/processing and the presence of permitted aluminium-containing additives.[12] Consequently, there is a wide range of dietary intakes of aluminium. Breast-fed infants can receive as little as 3-5μg day^{-1}, whereas infants fed specialised formulas can receive up to 500-2000μg day.$^{-1}$ [13,14] The aluminium content of some commercial baby foods can exceed 10mg day^{-1} and thus provide intakes of more than 2mg day^{-1}, which is comparable with the adult daily intake of 2-5mg day^{-1}.[15,16]

Small increases in plasma aluminium concentrations have been observed in infants fed soy formula compared with those fed whey-based formulae.[17,18] These observations which suggest that plasma aluminium concentrations might be a reasonable biomarker for aluminium intake, in the absence of overt toxicity, are supported by a recent collaborative study between Institute of Child Health London and University of Southampton.[19] Data from this study are summarised here.

3.1   Institute of Child Health/Southampton bioavailability study

3.1.1 Patients and Methods. Plasma and samples of feed were collected from 74 infants aged 14-112 days receiving a variety of feeds and who were hospitalised or attending routine postnatal outpatients clinics. All had normal renal function: none were uraemic. Healthly infants were fed breast milk or ready-to-feed whey based infant formulae. Preterm, fortified, soy, and a hydrolysed protein formulae (casein hydrolysate) were given to sick infants for disorders such as lactose intolerance, metabolic disorders, metabolic liver disease, gut surgery. Fortified milk was made from whey-based infant formula with glucose polymer and fat added. Infants had been established on a feed for a minimum of 2 weeks before the study. None had received solids or medications known to contain aluminium.

Stringent precautions were taken to minimise aluminium contamination. Breast milk was hand-expressed directly into low-aluminium-content bottles. Milk formulae were either ready-to-feed or reconstituted in the hospital milk room following manufacturer's directions and using deionised water which was shown to contain less than $0.5\mu g$ l^{-1} aluminium. Blood was collected by venepuncture between feeds to avoid temporal increases in plasma aluminium.[20] Low-aluminium-content needles and syringes, and lithium heparin tubes were used. Blood was centrifuged, and then separated using acid-washed Pasteur pipettes.

All samples were analysed by ETA-AAS using methods previously described.[20,21] Milk samples were digested with HNO_3 prior to analysis. Plasma samples were diluted $1+3$ with water followed by *in situ* O_2 ashing during ETA-AAS. The detection limit was $0.5\mu g$ l^{-1}. The within-and between-run precision was 5 to 6% over the range 8 to $160\mu g$ l^{-1}. Details of the internal quality control (IQC) of analysis and of external quality assessment (EQA) of analytical accuracy have been reported.[19]

3.1.2 Results. There was a variation of almost 200-fold between the lowest aluminium concentration in breast milk of 5.6μg l^{-1} and the highest, 914μg l^{-1}, in the casein hydrolysate (Table 2). The increases in aluminium in feeds were accompanied by increased plasma aluminium but statistical significance was only reached for the difference in plasma aluminium between the breast-fed infants and those receiving casein hydrolysate. It is possible that the small numbers of subjects may be partly responsible. Although the infants in this study consumed less than the normal weekly tolerable intake of $7000\mu g$ kg^{-1} body weight (the maximum was $932\mu g$ kg^{-1} bodyweight), those fed soy or casein hydrolysate formulae consumed 20 to 30 times more aluminium than did breast-fed infants. Raised plasma aluminium concentrations were seen in most infants who consumed feeds containing more than $300\mu g$ l^{-1}. There was a significant ($p=0.002$) relationship between the mean concentrations of aluminium in plasma and in the feeds (Figure 3). The regression equation indicates that the concentration of aluminium in plasma is increased by $8.5\mu g$ l^{-1} for an increase in the aluminium concentration of $1000\mu g$ l^{-1} in an infant's feed. These data which indicate plasma aluminium concentrations to be useful biomarkers of intake and uptake of aluminium from infant feeds are supported by balance studies which show substantial uptakes and retentions of aluminium by infants with high aluminium intakes.

Table 2   Subjects, types of feed and aluminium concentrations

Feed	No. of Infants	Aluminium*/μg l^{-1}, in,	
		Milk	Plasma
Breast	15	9.2(5.6-12.7)	8.6 (5.6-10.6)
Whey-based	24	165 (151-180)	9.2 (7.4-11.0)
Fortified	14	161 (143-180)	10.3 (8.3-12.3)
Pre-term	7	300 (272-280)	9.7 (5.3-17.1)
Soy	7	534 (470-598)	12.5 (5.0-20.0)
Hydrolysate	7	773 (632-914)	15.2 (10.7-19.8)

*DATA ARE MEANS (95%CI)

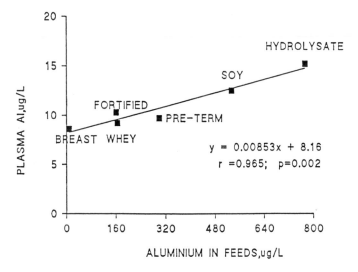

Figure 3   Mean concentrations of aluminium in plasma samples and feeds of 74 infants aged 14–112 days.

4 MERCURY

A study of mercury pollution as a consequence of gold-mining in the Brazilian Amazon showed mercury levels in blood and urine high enough to produce clinical symptoms of mercury poisoning in 13 of 106 individuals studied.[23] A particular cause for concern was the high level of mercury in the blood of villagers eating fish caught locally. The mercury concentration in fish ranged from $9\mu g$ kg[-1] up to $2,600\mu g$ kg[-1] fresh weight. Production in this area has been estimated at 12-20 tonnes of gold annually since 1979, with a similar discharge of mercury into the regional ecosystem. The city of Itaituba (population about 100,000) is the most important gold trading site in Amazonia. The other locations sampled were the gold-mining camps of Crepurí (population about 3,000) and Cuiú-Cuiú population about 1,000), and the fishing village of Jacareacanga on the river Tapajós, 100km from the nearest mining camps. Blood and urine samples were collected by a physician at improvised clinics, frozen within 48 hours, and mercury was determined at Southampton using inductively coupled plasma source mass spectrometry (ICP-MS) with stringent quality-control procedures.

A comparison of blood and urine mercury levels in subjects from Crepurí - the larger gold-mining camp and from the village at Jacareacanga yields fascinating data on the relative exposure to inorganic mercury and to methylmercury (Figures 4,5). Mercury in urine is an appropriate indication of the former whereas blood mercury can act as a biomarker for the ingestion of methylated (organic) mercury in fish tissue. The subjects from Crepurí, who would be exposed to inorganic mercury show marked elevations in mercury in urine (Figure 4) with 6 individuals having $\geq$ $100\mu g$ l[-1] a level at which there is a high probablility of developing classical signs of mercurialism. Their blood mercury levels were also elevated with 12 individuals (out of 25) having concentrations $\geq$ $20\mu g$ l[-1] which is the upper 95% limit for industrial workers in the UK.

The distribution of blood and urine mercury levels in the 25 villagers from Jacareacanga (Figure 5) is markedly different from that of the gold-miners. Only 1 subject had a urine mercury above $100\mu g$ l[-1] and he burned 1 kg Au (equivalent to 1kg Hg) per week. All of the other villagers studied had urine mercury concentrations below $35\mu g$ l[-1] However the incidence of elevated blood mercury levels of these villagers was much higher than the gold miners. Most, 80%, of the villagers studied had blood-mercury concentrations $\geq 20\mu g$ l[-1] *ie.* more than the upper limit for industrial workers, and 8 out of 25 had blood mercury levels $\geq 100\mu g$ l[-1]. Four subjects had blood mercury levels which either exceeded or approached the level of $200\mu g$ l[-1], which may be associated with neurological changes characterised by parasthesia.

The above data show clearly that the concentration of mercury in urine is a useful biomarker for exposure to inorganic (Hg^o) mercury (Figure 4) and that the concentration of mercury in whole-blood is a good biomarker for the intake and uptake of methylmercury from foods (Figure 5). Further evidence of the intake of methyl mercury from fish and of its reflection in blood mercury levels was obtained from a simple questionnaire on the consumption of fish by the villagers in Jacareacanga. The data, summarised in Figure 6, show that increased fish consumption from "1 to 2 days" per week to "every day" is associated with increased levels of mercury in blood.

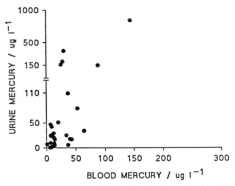

Figure 4   Mercury in blood and urine of gold-miners.
Crepuri Brazil

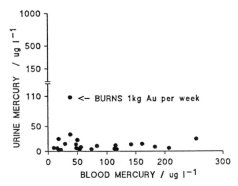

Figure 5   Mercury in blood and urine of fish-eaters.
Jacareacanga Brazil

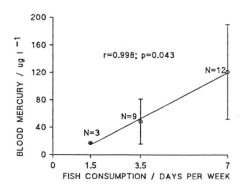

Figure 6   Effect of fish consumption on blood mercury
Jacareacanga Brazil

5   LEAD

Blood-lead concentrations provide valid estimates of recent exposure and uptake. Additional information on the relative importance of the various sources of lead exposure may be obtained from comparing the lead isotopic compositions of potential sources with those of the body fluids and tissues of the exposed subjects.[24] One example of the use of lead isotope ratios as biomarkers of lead uptake is a recent study of lead in drinking water in Blackburn UK.[25]

5.1   Lead in drinking water in Blackburn.

This was a small study established to investigate the possible risk to young children arising from contamination of drinking water by lead. The local water company, North West Water, had received many complaints annually from householders receiving highly discoloured water. In many of these samples the concentrations of lead exceeded $50 \mu g$ l^{-1} and in some cases were 10 to 1000 times greater than this concentration. Local agencies, the Department of Health and the Department of the Environment thought that the very high concentrations were likely to be intermittent and could not be used to estimate longer term exposure or predict consequences for health. It was also considered possible that consumers faced with highly discoloured water would not drink it and thus avoid exposure to high concentrations of lead. Nevertheless, it was considered desirable to explore these assumptions by trying to obtain blood samples for total and isotopic lead analysis from individuals in households which had high water lead concentrations. A small study was therefore undertaken jointly by Environmental Geochemistry Research Centre at Imperial College, London, and the Trace Element Unit at the University of Southampton.

5.1.1 Samples and methods. The sampling strategy and analytical protocol is detailed elsewhere.[25] Briefly, the study was limited to households in which there was a complaint about discoloured or cloudy water and in which the concentration of lead in water lead exceeded $100 \mu g$ l^{-1}. Individuals in households so identified were invited to provide blood samples for total and isotopic lead analysis. A total of 55 households provided water samples which were analysed for total lead. From these, a subset of 30 water samples were analysed for $^{206}Pb:^{207}Pb$ ratios. The provision of blood samples, however, was poor. Only 24 subjects from 10 households provided blood samples. There were 11 blood samples from 8 children (3 repeats) and 17 blood samples from 16 adults (1 repeat). In addition, the time interval between identification of the household and obtaining blood samples was highly variable and ranged up to 10 months. Samples of house-dust and garden soil were obtained from 3 of the 7 households which had provided both blood samples and water samples.

Particular matter collected on filters from a water pipe was analysed by scanning electron microscopy[26] at Imperial College, where in addition, concentrations of lead in drinking water, house-dust and garden-soil were measured by ICP-AES.[27] Total concentrations of lead in blood were measured by microsampling flame AAS[28] at Southampton where $^{206}Pb:^{207}Pb$ ratios were measured in blood, water, dust and soil by ICP-MS.[29] The accuracy of all analytical procedures was ensured by employing established internal quality control protocols.[28, 29, 30,31]

5.1.2 Results. The concentrations of lead in the water samples varied from 0.3 to 1236μg l^{-1} and within a given household day-to-day changes of up to 100 fold were not uncommon. Two reservoirs supplying the drinking water to this area had lead concentrations of only 0.24, and 0.47μg l^{-1}. The blood-lead concentrations ranged from 1.2 to 31.2 μg dl^{-1} with elevated concentrations ($\geq$ 10μg dl^{-1}) seen in 5 samples from 4 children and in 7 samples from 6 adults. The expected frequency of subjects with a blood lead > 10μg dl^{-1} is 1 in 20; in this study it was 10 in 24, indicating a greater than normal intake of lead.

The ^{206}Pb:^{207}Pb ratios in blood, drinking water, dust and soil are given for 10 households in Figure 7. The broken horizontal lines mark the limits of the average ratios expected for UK subjects with low lead exposure.[24] It can be seen that most of the blood samples from the Blackburn subjects had ^{206}Pb:^{207}Pb ratios greater than 1.13, *ie.* above the upper limit of the average range expected for UK subjects with low lead exposure. These data suggest an even higher ratio in the dominant source of lead. Isotopic analyses of the samples of garden-soils and house-dusts available from 3 households showed that these could not have been significant contributors to body lead because the ^{206}Pb:^{207}Pb ratios were lower than the ratios seen in the relevant blood samples. Only drinking water had ^{206}Pb:^{207}Pb ratios sufficiently high to have produced the increase in ratios of the blood samples above the expected range of 1.12 to 1.13. Waters from the reservoirs serving the area have low ratios as well as low total lead concentrations, but the water distribution pipes had very high ratios. Clearly dissolution of the lead piping by the drinking water produces a high lead concentration plus a change in ^{206}Pb:^{207}Pb ratio towards that of the piping. Increased uptake of this source of lead is shown by the increase in ratio in the blood of the Blackburn subjects as well as an increase in the blood-lead concentration.

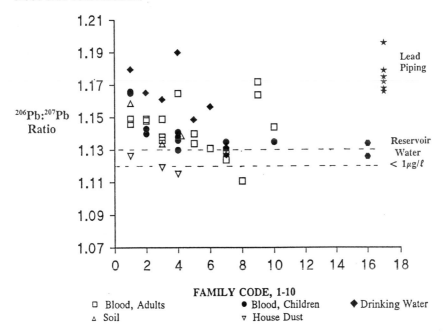

Figure 7 ^{206}Pb:^{207}Pb Ratios in Blood and In Environmental Samples from Blackburn

The relationship between the mean ^{206}Pb:^{207}Pb ratios in blood and in drinking water for the seven families for whom these samples were available was statistically significant (Figure 8). Given the long and variable time intervals between collection of the waters and of the blood samples this is particularly a striking observation.

Despite the paucity of samples in this study, the use of total and isotopic lead measurements as biomarkers of lead intake and uptake has indicated that in Blackburn the most likely source of lead in drinking water is contamination from communication/common service pipes and from pipes within the curtilage of individuals households and that lead in drinking water dominates exposure of the affected population.

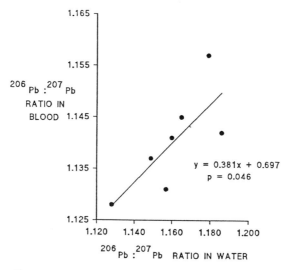

Figure 8   Comparison of mean lead isotope ratios in blood and
in drinking water for seven households in Blackburn.

6. CONCLUSION

The examples given have shown that concentrations of trace elements in body tissues and fluids can provide useful indications of intakes from foods. There is however scope for improvement. More information could be obtained from a study of the elemental species within a given specimen   rather than just the total concentration of all species. Concommitant species in the food can also influence elemental uptake *e.g.* citrate is known to enhance aluminium absorption. The presence of citrate in casein hydrolysate may be partly responsible for elevated plasma aluminium concentrations in infants receiving this feed (Table 2). Changes in lead isotope ratios in blood in response to changes in uptake may be influenced by the contribution of skeletal-lead to blood-lead, particularly for adults. An improved detection of these changes can be obtained by measuring all four stable isotopes of lead and not just ^{206}Pb:^{207}Pb ratios.

Notwithstanding these limitations, trace-elemental (and stable isotopic) concentrations in body fluids and tissues can be useful biomarkers of excessive intakes and uptakes from foods and beverages.

References

1. M. Yamada, S. Ohno, I Okayasu, R. Okeda, S. Hatakeyama, H. Watanabe, K. Ushio, H. Tsukagashi, *Acta Neuropathol.*, 1986, **70**, 273.

2. C-C. Huang, N-S. Chu, C-S.Hu, J-D. Wang, J-L Tsai, J-L. Tzeng, E.C. Wolters, D.B. Calne, *Arch. Neurol.*, 1989, **46**, 1104.

3. R.C. Baselt, R.H. Cravey, 'Disposition of toxic drugs and chemicals in man', 3rd Edn., Year book Medical Publishers, Chicago, 1989.

4. A. Ejima, T. Imamura, S. Nakamura, H. Saito, K. Matsumoto, and S. Momono, *Lancet.*, 1992, 359, 426.

5. J. Cawte and M. Florence, *Lancet*, 1987, **1**, 1484.

6. D.J. Halls and G.S. Fell, *Anal Chim Acta.,* 1981, **129**, 205.

7. H.T. Delves, in A Taylor, ed. 'Trace elements in human disease'. W.B. Saunders, London, 1985.

8. Department of Health, Medicine Controls Agency, Drug Alert Ref. EL(94)(ALERT) A/16, May 1994.

9. D.J. Halls in R.F.M. Herber and M. Stoeppler, eds 'Trace Element Analysis in Biological Specimens', Elsevier, Amsterdam, 1994.

10. D.B. Milne, R.L. Sims, and N.V.C. Ralston, *Clin. Chem.*, 1990, **36**, 450.

11. I.L. Shuttler, 'Investigations into the use of L'vov Platforms in ETA-AAS for the determination of trace elements in body tissues and fluids' PhD Thesis, University of Southampton 1988.

12. H.T. Delves, B. Suchak, and C.S. Fellows, in R.C. Massey and D. Taylor, eds, 'Aluminium in food and the environment', Royal Society of Chemistry, London 1989.

13. M.E. McGraw, N. Bishop, R. Jamison et al *Lancet.*, 1986, **1**, 157.

14. J.L. Greger in D.J. Chadwick and J. Whelan, eds. 'Aluminium in biology and medicine' Wiley and Sons, Chichester 1992.

15. B. Suchak., 'Determination of aluminium in foods and beverages' MPhil Thesis, University of Southampton, 1992.

16. H.T. Delves, C.E. Sieniawaska, and B Suchak., *Analyt Proc.*, 1993, **30**, 358.

17. R. Litov, V.S. Sickles, G.M. Chan, M.A. Springer and A.C. Cordano, *Pediatrics.*, 1989, **84**, 1105.

18. P. Goyens, and D. Brasseur, *Pediatrics.*, 1990, **86**, 650.

19. N.M. Hawkins, S. Coffey, M.S. Lawson and H.T. Delves, *J. Paediatr. Gastroent and Nutr.*, 1994, **19**, 377.

20. C.E. Sieniawska, 'Bioavailability of aluminium from foods and beverages' MPhil thesis, University of Southampton, 1993.

21. C.S. Fellows, 'Aluminium analysis in clinical samples by GF-AAS and some fundamental studies of ICP-MS' PhD Thesis, University of Southampton 1993.

22. M.S. Lawson and H.T. Delves, 'Preliminary and unpublished data from ongoing study.

23. D. Cleary, I. Thornton, N. Brown, G. Kazantzis, T. Delves and S. Worthington, *Nature.*, 1994, **369**.

24. H.T. Delves and M.J. Campbell, *Env.Geochem and Health.*, 1993, **15**, 75.

25. Department of Environment 'Lead in drinking water in Blackburn'. Report of Collaborative study, December 1994.

26. J.M. Watt, *Microsc. Anal.*, 1990, **15**, 25.

27. M. Thompson and J.N. Walsh, 'Handbook of inductively coupled plasma spectrometry', 2nd Edn, Blackie, Glasgow, 1989.
28. H T Delves, *Analyst., 1970, 95, 431.*
29. M.J. Campbell and H.T. Delves, *J. Analyt. Atom. Spectro.*, 1989, **4,** 235.
30. L.M. Alexander, A. Heaven and J. Moreton, *Arch. Environ. Health.*, 1993, **48,** 392.
31. Department of Environment, UK Blood-lead monitoring programme 1984-1987. Results for 1984. Pollution Report No. 22 HMSO.

Urinary Monitoring of Saccharin and Acesulfame-K as Biomarkers of Intake

L. A. Wilson and H. M. Crews

CSL FOOD SCIENCE LABORATORY, NORWICH RESEARCH PARK, COLNEY, NORWICH
NR4 7UQ, UK

1. INTRODUCTION

Reliable data on the use and consumption of food additives are important prerequisites for the development of good food legislation and safety policies. Data relating to the consumption of food additives can be obtained from several methods, including 24 hour recall and food frequency questionnaires and estimates based on the intake by the whole population calculated from food sales. There are, of course, advantages and disadvantages involved in these methods. They are non-invasive and relatively easy to carry out, but they can be unreliable and inaccurate.

European Union member states are being asked to establish a system of consumer surveys to monitor additive consumption [1-3]. A more prescriptive approach will be taken than is done so at present and better intake data will be required. Therefore an additional approach to intake data collection needs to be developed which would be accurate, sensitive, specific and simple to implement. To this end, the use of urinary biomarkers as a measure of consumption of some food additives has been considered. Biomarkers are defined as 'cellular, biochemical or molecular alterations which are measurable in biological media such as human tissues, cells or fluids and are indicative of exposure to environmental chemicals' [4]. They may be specific to a compound or group of compounds and may permit an assessment of exposure to those compounds.

Two artificial sweeteners, saccharin and acesulfame-K, have been identified as being suitable candidates for the biomarker approach. They are both widely used by food and pharmaceutical manufacturers, and are consumed at high levels, especially by certain population groups, e.g. children, diabetics and slimmers (ca. 35 - 50% of the Acceptable Daily Intake). Both compounds pass unmetabolised through the body to be excreted in the urine [5], making them ideal additives for this study.

The first objective of the study was to develop a suitable method of analysis for the sweeteners in urine. Secondly, a pilot study with five volunteers was used to investigate whether an intake - excretion (dose - response) relationship could be established by controlled dosing of the volunteers with low, medium and high levels of the sweeteners. This pilot study also enabled the sampling and method protocols to be checked and revised as necessary. A controlled intake study on a larger group of volunteers for a longer period of time will follow to check the robustness of both the revised methodology and the dose - response relationships. Finally, if this last study is successful, a random population (and specific population groups, e.g. children) will be studied using both 24 hour urine collections and dietary questionnaires. This paper reports the results from the first two stages of this work.

2. METHODOLOGY

Methods for the measurement of acesulfame-K in foods[6] and saccharin in urine[5] were adapted to allow the extraction, clean-up, concentration and analysis of urine samples.

2.1. Materials

All reagents and solvents were HPLC grade or better. Millipore grade water was used throughout. Sodium saccharin was purchased from Sigma Chemicals Co. and acesulfame-K was kindly donated by Hoechst (UK) Ltd.

2.2. Extraction of bulked 24 h urine samples

Each urine sample was shaken well to ensure homogeneity. Sub-samples (2 mL) were pipetted into 20 mL separating funnels and 0.1 M potassium dihydrogen orthophosphate (pH 7.0) (1 mL) added. The sample was acidified by the addition of concentrated sulphuric acid (200 µL). Diethyl ether (10 mL) was then immediately added. The separating funnel was shaken vigorously for 60 seconds and the phases allowed to separate. The organic layer was removed to a drying tube and a fresh aliquot of diethyl ether (10 mL) added to the separating funnel, which was shaken again for 60 seconds. The second organic layer was combined with the first in the drying tube and the solvent removed under a gentle stream of nitrogen, with heating at 20°C. The residue was dissolved in 0.1 M potassium sulphate (500 µL).

2.3. Clean-up

Each sample extract was filtered through a 0.2 µm syringe filter (Anachem, Luton, UK) before injection onto an ion-chromatography (IC) system. This consisted of a 100 µL loop, and a 4 x 250 mm IonPac AS4A column (Dionex (UK) Ltd, Camberley, UK). A guard column of the same chemistry was also fitted. Detection was by UV at 225 nm. 0.1 M potassium sulphate was used as the mobile phase with a flow rate of 2 mL/min. A bulked urine sample extract fortified with sodium saccharin and acesulfame-K (at about 15 µg/mL (saccharin) and 30 µg/mL (acesulfame-K)) was injected before each batch of samples to check the retention times of the sweeteners. The clean-up was effected by collecting the fraction eluting from the detector at the appropriate time, i.e. usually between about 2 to 4 minutes.

2.4. Concentration

The collected fraction was acidified with concentrated sulphuric acid (100 µL) and extracted immediately with diethyl ether (ca. 5 mL). The organic layer was removed to a drying tube and the extraction repeated with a second aliquot of diethyl ether. The second organic layer was added to the first and the combined extracts dried down under a gentle stream of nitrogen with heating to 20°C. The residue was dissolved in analytical HPLC mobile phase (400 μL).

2.5. Analysis and quantification

Samples were quantified by injection (30 μL) onto a 4.6 x 250 mm Kromasil C18 column (Thames Chromatography, Maidenhead, UK), fitted with a guard column of the same chemistry. Detection was by UV at 225 nm. The mobile phase was 20 mM ammonium acetate (pH 3.5) : acetonitrile (9 : 1) with a flow rate of 0.7 mL/min. The limit of detection (LOD) was established as 0.02 μg/mL for both compounds and the limit of quantification (LOQ) as being 0.07 μg/mL. A set of calibration standards, ranging from ca. 0.45 - 30 μg/mL (saccharin) and ca. 1 - 60 μg/mL (acesulfame-K), were run prior to each batch of analyses. Quantification was achieved by constructing calibration curves and calculating levels of the sweeteners by interpolation of the relevant graph. The curves for both analytes were linear over the range studied. Results are quoted as saccharin and acesulfame-K.

2.6. Quality assurance

Each batch of samples consisted of no more than 4 samples extracted in duplicate. One of the samples (usually a blank bulk sample) was also extracted after fortification with a sodium saccharin / acesulfame-K spiking mix. The same sample, unfortified, was extracted to act as the spike 'blank'. Usually, 2 spiked samples at different levels were extracted with each batch, making a maximum batch size of 10 samples. Reagent blanks were also extracted to ensure that no interfering co-extractives were present.

2.7. Method validation

The extraction method was validated by spiking urine samples with known amounts of sodium saccharin and acesulfame-K and extracting them as above (2.2.). The recoveries are reported in Table 1. Sample extracts were also analysed by HPLC-MS to confirm peak identities. The chromatographic conditions were essentially the same (the flow rate was increased to 1 mL/min). Atmospheric pressure chemical ionisation (APcI) was used, which was found to give better linearity for both compounds than other ionisation techniques tried.

2.8. Pilot study sample collection and storage

Each volunteer was given a 1000 mL measuring jug and sufficient 500 mL bottles for 24 hours. Each urine sample was collected into a sample bottle and labelled with the volunteer's name and the date and time the sample was taken. Upon receipt at the laboratory, the samples were weighed to allow calculation of the sample volume (using the mean density of urine, taken as 1.015 g/mL [7]), and 10% (v/v) of each individual sample taken and combined to make a bulk sample. The bulk and remaining individual samples were frozen at -25°C until required for analysis. When required, the samples were allowed to defrost overnight at room temperature before sub-sampling or over a weekend at +4°C.

3. PILOT STUDY

3.1. Subjects

Five volunteers agreed to participate in the pilot study, 2 adult males (aged 35 - 37 years; weight 70 - 74 kg) and 3 females (aged 25 - 48 years; weight 50 - 69 kg). The subjects were given oral doses of the sweeteners in the form of retail sweetener tablets (Boot's own brand saccharin tablets and Boot's own brand 'Gold' acesulfame-K tablets) dissolved in water. The subjects refrained from consuming saccharin and acesulfame-K for 1 week before the first dose and then throughout the study. The course of the study is shown below:

DAY 1   Saccharin and acesulfame-K free diet begins

DAY 8   Saccharin and acesulfame-K free diet continues
             24 hour urine collection : BLANK

DAY 9   **Low dose**
             24 hour urine collection : LOW

DAY 10   **Medium dose**
             24 hour urine collection : MEDIUM

DAY 11   **High dose**
             24 hour urine collection : HIGH

3.2. 'Dose' preparation

The doses were prepared by dissolving the sweetener tablets in water. The concentration of the sweeteners in the tablets was established by dissolving a single tablet of each type in 1000 mL of water and injecting the solution onto the analytical system. Quantification was carried out by the use of calibration curves. Using this method, the saccharin tablets were found to contain 12.60 mg of sweetener and the acesulfame-K tablets 20.90 mg of sweetener per tablet. The manufacturers of the tablets were contacted and confirmed that the saccharin concentration should be 12.5 mg/tablet and the acesulfame-K concentration should be 20 mg/tablet.

3.2.1. Low dose. One of each type of tablet was dissolved in 1000 mL of water. Each volunteer drank 100 mL of this solution (equivalent to 1.26 mg of saccharin and 2.09 mg of acesulfame-K).

3.2.2. Medium dose. One of each type of tablet was dissolved in water (ca. 500 mL) and drunk. The dosing cups were rinsed out twice with water or sweetener-free soft drink and the rinsings also drunk (equivalent to 12.6 mg of saccharin and 20.9 mg of acesulfame-K). The soft drink was analysed prior to the study by degassing by sonication, then quantification by direct injection onto the analytical HPLC system. No peaks corresponding to the retention times of saccharin or acesulfame-K were detected.

3.2.3. High dose. Five of each tablet were dissolved in water (ca. 500 mL) and drunk. The dosing cups were rinsed out several times with water or sweetener-free soft drink and the rinsings also drunk (equivalent to 63.0 mg of saccharin and 104.5 mg of acesulfame-K).

In summary, the doses can be related to Acceptable Daily Intake (ADI) levels for the sweeteners as follows, where the ADI = 0 - 5 mg/kg body weight/day for saccharin and the ADI = 0 - 9 mg/kg body weight/day for acesulfame-K.

Table 1: Sweetener doses and their relationship to the ADI

Dose	Saccharin		Acesulfame-K	
	mg	[a]normalised	mg	[a]normalised
Low (L)	1.26	0.02	2.09	0.04
Medium M)	12.60	0.21	20.90	0.35
High (H)	63.00	1.05	104.50	1.74

Key: [a] = normalised to ADI and mg/kg body weight, where 60 kg = body weight

4. RESULTS

4.1. Method validation

To validate the extraction method, sub-samples of urine (2 mL) were fortified with known concentrations of sodium saccharin and acesulfame-K and extracted and quantified. The results (Table 2) are shown below.

Table 2. Recoveries of saccharin and acesulfame-K

Saccharin spike concentration (μg/mL)	Recovery (%)	Acesulfame-K spike concentration (μg/mL)	Recovery (%)
0.45	81.1	0.99	70.7
2.21	104.3	4.96	72.0
4.44	97.3	9.91	79.6
6.65	87.8	14.87	68.2
13.31	72.5	29.74	60.2
26.61	73.5	59.47	60.5
Mean	86.1	Mean	68.5
RSD	15.0	RSD	10.8

4.2. Quality Assurance

All sample batches consisted of urine samples (extracted in duplicate) and spiked urines fortified at different levels. The recoveries from these spikes are given in Table 3.

Table 3. Recoveries of saccharin and acesulfame-K from spiked urines

Saccharin spike concentration (µg/mL)	Mean recovery (%)	Acesulfame-K spike concentration (µg/mL)	Mean recovery (%)
2.18	71.1	5.00	71.0
4.37	70.3	9.95	70.7
4.44	82.0	10.00	69.0
6.56	72.3	14.92	70.7
6.65	81.3	15.00	72.3
13.11	74.3	29.74	75.9
13.31	96.5	30.00	51.4
26.23	90.0	59.47	92.0
26.61	102.5	60.00	67.2
Mean	81.1	Mean	69.0
RSD	14.7	RSD	17.3

Recoveries are means of duplicate extractions.

4.3. Creatinine analysis

The rate of excretion of creatinine from an individual's body is reported as being relatively constant [8]. Each bulk urine sample was analysed to determine creatinine concentration so that if necessary a normalisation could be carried out. After consideration, it was felt that normalisation by creatinine levels would not be the most accurate method to use, especially in the light of other work [9].

4.4 Pilot study results for 24 h urine samples

Each subject's bulked 24 hour urine samples were analysed as described in (2.2). The results were corrected for recovery according to the appropriate sample spike (see Table 3) and expressed as sweetener excreted (mg). See Tables 4 and 5. The amount of sweetener excreted expressed as a percentage of the dose administered is also shown, see Table 6. The results detailed in Tables 4 and 5 are also presented graphically in Figures 1 and 2 to facilitate comparisons between subjects and sweeteners.

Table 4. Saccharin excreted in 24 h urine samples

Subject	Dose (mg)			
	0	1.26	12.60	63.00
1	nd	0.69	11.03	30.68
2	nd	0.49	11.44	45.26
3	1.87	1.56	10.07	66.22
4	0.42	1.01	10.32	45.42
5	2.40	2.80	14.10	46.80

Key: nd = not detected

Table 5. Acesulfame-K excreted in 24h urine samples

Subject	Dose (mg)			
	0	2.09	20.90	104.50
1	nd	1.53	22.78	107.11
2	11.22	5.48	18.42	99.85
3	nd	1.15	16.32	105.18
4	nd	1.52	17.30	88.07
5	0.22	3.23	29.01	98.63

Key: nd = not detected

Table 6. Sweeteners excreted expressed as percentage of dose

Subject	Dose (see Table 1)	% saccharin dose found in urine	% acesulfame-K dose found in urine
1	L	54.8	73.2
	M	87.5	109.0
	H	48.7	102.5
2	L	38.9	262.2
	M	90.2	88.1
	H	71.8	95.6
3	L	123.8	55.0
	M	79.9	78.1
	H	105.1	100.7
4	L	80.2	72.7
	M	81.9	82.8
	H	72.1	84.3
5	L	222.2	154.5
	M	[a]111.9	[a]138.8
	H	74.3	94.4

Key: nd = not detected
      B = blank sample (from Day 8)
      L = low dose collection (Day 9)
      M = medium dose collection (Day 10)
      H = high dose collection (Day 11)
      [a] = 1 individual sample not collected

Figure 1. Pilot Study: Saccharin results for 5 volunteers

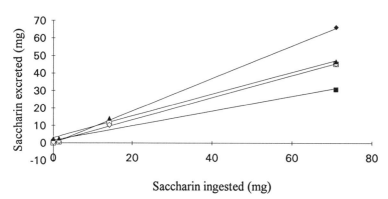

Saccharin ingested (mg)

Figure 2. Pilot Study: Acesulfame-K results for 5 volunteers

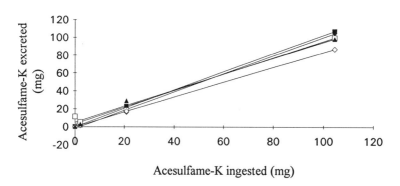

Acesulfame-K ingested (mg)

5. DISCUSSION AND CONCLUSIONS

As a result of the pilot study, the methods for extraction and measurement of 24 hour urine samples reported here have recently been improved by revising the extraction stage, and mean recoveries of 99.3% and 86.5% for saccharin and acesulfame-K were obtained. The improved method, (L.A. Wilson, 1995, unpublished), will be used for all future work.

However, the results obtained for the five person pilot study have demonstrated that an intake versus excretion response is possible using the methodology described. Figures 1 and 2 indicate that the variation between volunteers was less for acesulfame-K than for saccharin. One problem with saccharin is that it occurs in commodities other than food. It has been suggested that one such source is toothpaste [9]. Investigations showed that 3 of the brands used by our volunteers contained saccharin. The fifth brand was not labelled, but appeared to contain acesulfame-K rather than saccharin (subject 2). The only subject not to have any contamination in their blank (subject 1) used baking soda rather than toothpaste for the duration of the study! This example serves to emphasise the need to define accurately what is included in the term 'dietary intake'.

A second example illustrates the need to obtain complete urine collections. For example, subject 5 missed one urine sample collection during the medium dose. As a result, the total 24 hour urine volume was reduced, leading to a slightly higher concentration of sweeteners in the urine for that dose. This problem of incomplete collections is one addressed by several other workers, notably Bingham *et al* [10] where a marker is ingested by the subjects at the beginning of the collection period and its concentration determined in the urine at the end of collection. The marker used in Bingham's study, *para*-aminobenzoic acid (PABA), has been obtained and evaluation of detection methods begun.

This paper reports data collected over single 24 hour periods. Future work will incorporate changes implemented as a result of the work described here including the use of three consecutive 24 hour collections to validate the methods and protocols.

The use of biomarkers of exposure to monitor dietary intake will not be a feasible approach for many additives. It is hoped that information obtained from those additives which are amenable to this approach will complement the use of dietary surveys which depend upon estimation and add to the accuracy of dietary intake data.

References

1. European Commission, *Directive on Sweeteners Used In Foodstuffs*, 94/35/EC, 1994

2. European Commission, *Directive on Colours Used in Foodstuffs*, 94/36/EC, 1994.

3. European Commission, *Directive on Food Additives Other Than Sweeteners and Colours*, 95/2/EC, 1995

4. H. S. Hulka, *Arch. Environ. Health*, 1988, **42**, 83

5. A. G. Renwick, *Xenobiotica*, 1986, **16**, (10/11), 1057

6. J. Prodolliet and M. Bruelhart, *J.A.O.A.C*, 1994, **76**, (2), 268

7. C. Lentner, Geigy Scientific Tables, **1**, CIBY-GEIGY, Basle, 1981

8. G. H. Bell, D. Emslie-Smith and C.R. Paterson, Textbook of Physiology, Churchill Livingstone, Edinburgh, London and New York , 1980

9. T.W. Sweatman, A.G. Renwick, and C.D. Burgess, *Xenobiotica*, 1981 **11**, (8), 531-540

10. S. Bingham, D.R.R. Williams, T.J. Cole, C.P. Price and J.H. Cummings. *Ann Clin Biochem*, 1988, **25**, 610

The Use of the Comet Assay as a Biomarker to Measure the Genetic Effects of Food Chemicals and the Protective Effects of Antioxidants Administered *In Vitro* and *In Vivo*

D. Anderson,[1] Tian-Wei Yu,[1] B. J. Philips,[1] P. Schmezer,[2] A. J. Edwards,[1] K. Butterworth,[1] and R. Ayesh[1]

[1] BIBRA INTERNATIONAL, WOODMANSTERNE ROAD, CARSHALTON, SURREY SM5 4DS, UK

[2] GERMAN CANCER RESEARCH CENTRE, IM NEUENHEIMER FELD 280, 69120 HEIDELBERG, GERMANY

1 INTRODUCTION

A number of biomarkers of genetic damage, such as chromosome damage and sister chromatid exchange, have become well established as methods for monitoring human exposure and response to environmental mutagens[1]. Recently, the COMET assay has shown considerable promise as a rapid and sensitive method for detecting DNA damage[2]. It can be applied to a wide variety of cells which can be assessed immediately after isolation. The method is based on single cell gel electrophoresis. Chemically-induced breakage results in greater electrophoretic migration of DNA which can be quantified by image analysis after fluorescence staining. The involvement of DNA strand breaks in mutagenesis has been reviewed by Phillips and Morgan[3].

We have explored the potential of the COMET assay as a biomarker for assessing both the genetic damage caused by food mutagens and the degree of protection against such damage afforded by normal dietary components. The bulk of the work described here is concerned with the genetic effects of oxygen radicals and the protective effects of dietary antioxidants. It has long been known that oxygen, at concentrations greater than that in normal air, causes damage to tissues and whole organisms[4]. Many chemicals are known to generate DNA damage through oxygen radical mechanisms. As model compounds, we have used H_2O_2 and bleomycin, both of which have been shown to induce chromosomal aberrations, gene mutations and DNA strand breaks[2,5-9]. The effects of these compounds and the protective effects of a variety of antioxidant enzymes and chemicals have been examined using the COMET assay in human lymphocytes. The methods used and the results obtained in these *in vitro* experiments have been described elsewhere[10] but can be summarised as follows. H_2O_2 and bleomycin both produced clear dose-related responses in the COMET assay. The results were consistent throughout a total of 38 experiments with 2 lymphocyte donors. At low doses, vitamin C had a weak inhibitory effect on damage induced by H_2O_2 but at high concentrations was pro-oxidant and induced DNA damage in its own right. Water-soluble vitamin E (Trolox) had no effect but the plant flavonoid silymarin protected against H_2O_2. The iron-binding agents apo-transferrin and deferoxamine protected against bleomycin but not H_2O_2. Catalase was completely effective against H_2O_2 but superoxide dismutase was ineffective. Both enzymes gave a slight protection against bleomycin.

To test the usefulness of the COMET assay as a biomarker for the genetic effects of food chemicals in man, it has been included in an *in vivo* trial in which human volunteers have been given vitamin C supplements and their effect on oxidative damage assessed using a number of biomarkers. This trial was designed to take account also of the possible influence of high serum cholesterol levels on oxidative damage. It is still in progress but its design will be described below.

2 STUDY DESIGN

A group of 48 non-smoking volunteers was selected, from a panel of over 100, on the basis of plasma cholesterol levels. They were allocated to 3 groups of 16, each consisting of 4 males with low cholesterol levels matched for age and build with 4 males with high cholesterol levels, and 8 females matched in the same way. None of the volunteers was taking medication to control cholesterol levels and they maintained their normal dietary habits so as not to compromise cholesterol status. All procedures were performed to the standards of Good Clinical Practice. In this trial, a three-treatment, three period cross-over design (shown in Table 1) was adopted to take account of temporal differences in response as demonstrated for certain parameters by Dewdney *et al.*[11]

Prior to the main trial, a preliminary study was conducted to determine that the parameters to be examined and the methods to be used were suitable. Vitamin C was administered at a dose of 60 mg/day (recommended daily allowance) or 6 g/day for 14 days. No placebo was used. Blood samples were taken and the plasma tested for vitamin C level (by HPLC), cholesterol (enzymatically), total antioxidant capacity (ABTS method[12]) and lipid peroxidation products (LPO-586 method, Bioxytech S.A., France). Lymphocytes were separated and tested for DNA damage using the COMET assay, with and without challenge with H_2O_2. The results of the preliminary study only are reported.

3 RESULTS

3.1 Preliminary Vitamin C Supplementation Trial

A serum cholesterol cut-off point of 6 mmol/l was used to define subjects as "high" or "low" cholesterol. There were 12 high cholesterol males (mean 6.66 mmol/l before trial, 6.70 after the trial) and 12 low cholesterol males (mean 4.43 before trial, 5.13 after), 7 high cholesterol females (mean 6.67 before and 6.70 after) and 17 low cholesterol females (mean 4.12 before and 5.03 after). The serum cholesterol levels are shown in Figures 1 and 2. For most subjects, the classification of high or low cholesterol remained the same throughout the trial.

Table 1 *Human volunteer vitamin C supplementation trial - study plan*

Treatment:	Placebo	60 mg/day	6 g/day
WEEK:			
1+2	A	B	C
			BLOOD SAMPLE 1
3-8		WASHOUT PERIOD	
9+10	B	C	A
			BLOOD SAMPLE 2
11-16		WASHOUT PERIOD	
17+18	C	A	B
			BLOOD SAMPLE 3

A,B,C: Groups of 4 pairs of males and 4 pairs of females. Each pair matched for age and build but maintaining greatest possible difference in cholesterol level.

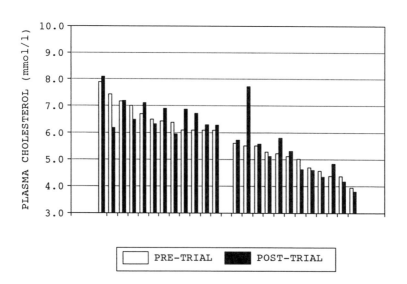

Figure 1 *Plasma cholesterol levels in male volunteers before and after vitamin C supplementation (preliminary study).*

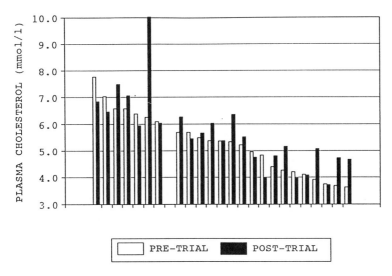

Figure 2 *Plasma cholesterol levels in female volunteers before and after vitamin C supplementation (preliminary study).*

Plasma vitamin C levels were measured after supplementation and the results are shown in Figure 3. The mean for the low dose group (60 mg/day) was 11.95 μg/ml and the high dose group (6 g/day) had a mean of 16.10 μg/ml.

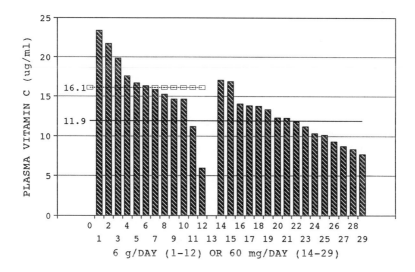

Figure 3 *Plasma vitamin C levels in volunteers after supplementation*

Plasma total antioxidant capacity values are shown in Table 2. The mean value for males on the high vitamin C dose increased by about 33%, that for females by only about 10%. Overall, there was a significant increase in antioxidant capacity at the higher vitamin C level (p < 0.01) and a marked difference between males and females (p < 0.001).

Table 2 *Plasma total antioxidant capacity; ABTS method - mean lag time (number of volunteers in parenthesis)*

Vitamin C	Cholesterol	Lag time (sec)	Mean
1) MALES			
HIGH	HIGH (6)	219.7	
	LOW (6)	193.5	206.6
LOW	HIGH (7)	150.0	
	LOW (6)	160.5	154.9
2) FEMALES			
HIGH	HIGH (3)	153.3	
	LOW (9)	127.3	133.8
LOW	HIGH (4)	137.3	
	LOW (7)	121.6	127.3
MEANS MALES:	179.7	VIT C 6g/DAY :	170.2
FEMALES:	130.7***	VIT C 60mg/DAY:	142.2**

Analysis of variance; significance obtained using least significant difference test:
*** = p < 0.01; *** = p < 0.001.*

Table 3 shows the results of COMET assays on freshly isolated lymphocytes either with or without treatment *in vitro* with H_2O_2. For clarity, only tail moment is shown. In both males and females, with and without H_2O_2, there were minor differences but no clear overall pattern.

3.2 Sensitivity of the COMET Assay to Food Mutagens

To demonstrate that the COMET assay was sensitive to food mutagens other than those acting by oxidative mechanisms, two heterocyclic aromatic amine mutagens were tested; PhiP (2-amino-1-methyl-6-phenylimidazo(4,5-b)pyridine) and IQ (2-amino-3-methylimidazo-(4,5-f) quinoline). As shown in Table 4, both gave dose-related effects in human lymphocytes in the presence of a metabolic activation system (S9).

Table 3 *Results of COMET assay on lymphocytes; Median values for tail moment (integrated value of density of DNA and migration distance) for untreated cells and cells treated with 10μM H_2O_2 (number of volunteers in parenthesis).*

Vitamin C	Cholesterol	Untreated	Treated
1) MALES			
LOW	ALL (13)	0.52	57.0
HIGH	ALL (8)	0.46^{ns}	54.2^{ns}
LOW	HIGH (7)	0.47	57.4
	LOW (6)	0.58^{**}	48.8^{ns}
HIGH	HIGH (4)	0.47	55.5
	LOW (4)	0.45^{ns}	53.2^{ns}
2) FEMALES			
LOW	ALL (11)	0.66	60.1
HIGH	ALL (8)	0.49^{*}	55.3^{ns}
LOW	HIGH (4)	0.76	60.4
	LOW (7)	0.57^{ns}	59.8^{ns}
HIGH	HIGH (2)	0.43	32.0
	LOW (6)	0.52^{ns}	60.5^{***}
3) MALES + FEMALES			
LOW	ALL (24)	0.56	58.2
HIGH	ALL (16)	0.47^{*}	54.5^{ns}
ALL MALES (21)		0.50	55.7
ALL FEMALES (19)		0.58^{ns}	57.9^{ns}

Analysis based on Mann-Whitney test: ^{ns} = *not significant;* * = $p < 0.05$; ** = $p < 0.01$; *** = $p < 0.001$.

4 DISCUSSION

Previous studies[2,10] have shown that the COMET assay is a rapid and sensitive method for measuring DNA damage in human cells. It has a number of advantages over other genetic biomarkers, such as chromosome analysis, since it does not require dividing cells and can be performed immediately after isolation of the cells. The number of mutagens tested has been extended with the demonstration here that food mutagens such as PhIP and IQ are positive in human lymphocytes.

Table 4 *Results of COMET assays in human lymphocytes treated with two food mutagens in the presence of rat liver S9.*

Treatment	Tail Moment (Median values)	
	Expt 1	Expt 2
Control	0.8	0.9
PhIP (μg/ml)		
2	0.5^{ns}	1.2^{ns}
10	4.2^{***}	3.0^{***}
50	11.5^{***}	8.7^{***}
IQ (μg/ml)		
5	0.7^{ns}	0.9^{ns}
25	1.1^{ns}	1.7^{*}
125	1.6^{**}	1.5^{*}

Median tail moment compared with control using Mann-Whitney test:
^{ns} = not significant; * = p < 0.05; ** = p < 0.01; *** = p < 0.001.
PhIP = 2-amino-1-methyl-6-phenylimidazo(4,5-b)pyridine; IQ = 2-amino-3-methylimidazo-(4,5-f) quinoline.

The potential of the COMET assay as an *in vivo* biomarker of genetic damage is under study and its performance will be compared with that of chromosome analysis in the ongoing main vitamin C trial. In the preliminary study, no firm conclusions can be drawn because only subjects receiving vitamin C supplementation have been examined (a placebo group was not included). Our previous study[10] showed that vitamin C at low concentrations had a protective effect against H_2O_2-induced damage in the COMET assay *in vitro*. It is to be anticipated, therefore, that the placebo group in the main trial might have a higher level of DNA damage, revealing the effect of vitamin C supplementation.

There was a 50% increase in the plasma vitamin C level in subjects receiving 6 g/day dose compared with those receiving 60 mg/day. There was a corresponding increase of up to 30% in the plasma antioxidant capacity in males and 10% in females. The higher vitamin C dose appears to have increased the antioxidant capacity more in the high cholesterol than in the low cholesterol group. There is a marked difference in the plasma antioxidant capacity of males and females. Differences between males and females have been encountered previously when using other biomarkers such as chromosome damage and SCE (e.g. Anderson *et al.*[13,14]).

References

1.   International Programme on Chemical Safety: Environmental Health Criteria **155**; WHO (1993).

2. V.J. McKelvey-Martin, M.H.L. Green, P. Schmezer, B.L. Pool-Zobel, M.P. Demec, & A. Collins, *Mutation Res.*, 1993, **288**, 47.

3. J.W. Phillips & W.F. Morgan, *Environ. Mol. Mutagen.*, 1993, **22**, 214.

4. B. Halliwell & J.M.C. Gutteridge, *Biochem. J.*, 1984, **219**, 1.

5. H. Joenje, *Mutation Res.*, 1989, **219**, 193.

6. M.L. Larramendy, M.S. Bianchi & J. Padron, *Mutation Res.*, 1989, **214**, 129.

7. L.F. Povirk & M.J.F. Austin, *Mutation Res.*, 1991, **257**, 127.

8. J. Rueff, A. Bras, L. Cristovoa, J. Mexia, M. Sa daCosta & V. Pires, *Mutation Res.*, 1993, **289**, 197.

9. K. Sankaranarayanan, *Mutation Res.*, 1991, **258**, 75.

10. D. Anderson, T-W. Yu, B.J. Phillips, & P. Schmezer, *Mutation Res.*, 1994, **307**, 261.

11. R.S. Dewdney, D.P. Lovell, P.C. Jenkinson & D. Anderson, *Mutation Res.*, 1986, **171**, 43.

12. N.J. Miller, C. Rice-Evans, M.J. Davies, V. Gopinathan & A. Milner, *Clinical Science*, 1993, **84**, 407.

13. D. Anderson, R.S. Dewdney, P.C. Jenkinson, D.P. Lovell, K. Butterworth & D.M. Conning, "Monitoring of occupational genotoxicants", M. Sorsa & H. Norppa, Liss, New York, 1986, p 39.

14. D. Anderson, P.C. Jenkinson, R.S. Dewdney, A.J. Francis, P. Godbert & K. Butterworth, *Mutation Res.*, 1988, **204**, 407.

Acknowledgements

We should like to thank D. Renstead for the statistical analysis. We are indebted to the Ministry of Agriculture, Fisheries and Food for funding the human volunteer study and the World Health Organisation for funding the *in vitro* studies.

Biomonitoring of Heterocyclic Aromatic Amines

R. J. Turesky

DEPARTMENT OF QUALITY ASSURANCE AND SAFETY EVALUATION, NESTEC LIMITED, RESEARCH CENTRE, VERS-CHEZ-LES-BLANC, 1000-LAUSANNE 26, SWITZERLAND

1 INTRODUCTION

Molecular biomarkers are an increasingly important tool for toxicologists and epidemiologists in assessing exposure and the adverse health effects posed by environmental and dietary contaminants[1-6]. The measurement of contaminants in the environment or food provides an estimate of exposure but this data is not sufficient to assess human health risk. Many contaminants, including heterocyclic aromatic amines (HAAs), require metabolism to exert their toxic effects. Therefore, it is necessary to measure absorption and formation of toxicologically active metabolites in target tissues to determine the biologically effective dose and assess risk. The capacity of individuals to metabolically activate or detoxify chemicals may vary greatly because of genetic polymorphisms[7]. Consequently, there may be individuals with elevated susceptibilities towards contaminants. In many instances, the toxic metabolites are unstable and their measurement is not possible. Therefore, chemically stable derivatives of these metabolites, such as their decomposition products, or macromolecular adduction products, are required to measure the biologically effective dose. Several of the biomarkers under development to assess exposure and the health risk of HAAs include urinary metabolites, protein and DNA adducts. The application and findings of these biomarkers in assessing human exposure are presented.

1.1 Heterocyclic Aromatic Amines: Formation, Exposure and Toxicity

More than a dozen heterocyclic aromatic amines are formed in tobacco, and in meats and fish cooked under typical household cooking practices[8,9]. Four chemicals which are structurally representative of this class of genotoxins are shown in Figure 1. Studies with model systems have shown that amino acids, sugar and creatinine are critical precursors in the formation of HAAs[10]. The HAA content in cooked foods vary and is dependent upon the meat and the method of preparaton[9,11]. Typical levels of HAAs found in grilled meat, fish and bacon range from 0.1 to 50 parts per billion[8,9,11].

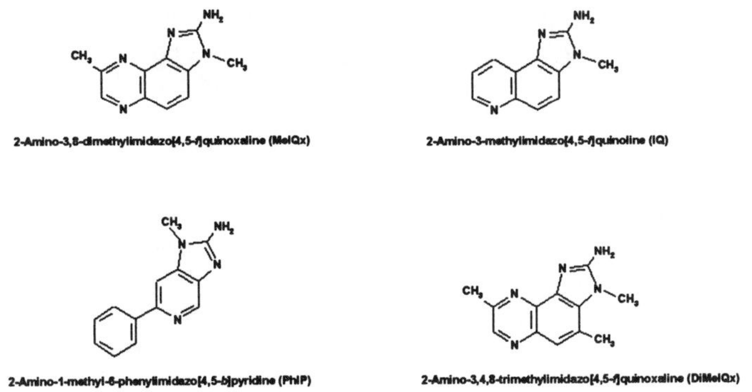

2-Amino-3,8-dimethylimidazo[4,5-*f*]quinoxaline (MeIQx)

2-Amino-3-methylimidazo[4,5-*f*]quinoline (IQ)

2-Amino-1-methyl-6-phenylimidazo[4,5-*b*]pyridine (PhIP)

2-Amino-3,4,8-trimethylimidazo[4,5-*f*]quinoxaline (DiMeIQx)

Figure 1 *Structures of several heterocyclic aromatic amines found in tobacco and cooked meats and fish*

1.1.1 *Genotoxicity of heterocyclic aromatic amines.* HAAs require metabolism to exert their genotoxic effects in bacteria and mammalian cells[8,9]. Metabolism to the carcinogenic *N*-hydroxy-HAA metabolites is catalysed by CYP1A2 in humans[12-14]. The *N*-hydroxy metabolites can readily bind to DNA or may be further activated following conjugation reactions, such as *N,O*-acetyltransferase or sulfotransferase, to form highly unstable esters which react with DNA (Figure 2)[14,15]. When given as part of the daily diet, HAAs induce tumors at multiple sites in mice and rodents[8,16]. Recently, 2-amino-3-methylimidazo[4,5-*f*]quinoline (IQ) has been reported to be a powerful liver carcinogen in the non-human primate[17]. Thus, because of the wide spread occurrence of these chemicals in our diet and the ability of human tissue to convert these chemicals to genotoxic species, there is concern that HAAs may be involved in the etiology of human cancers.

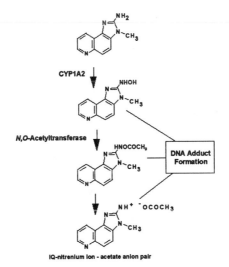

CYP1A2

N,O-Acetyltransferase

DNA Adduct Formation

IQ-nitrenium ion - acetate anion pair

Figure 2 *Metabolic activation of IQ through N-oxidation and N,O-acetyltransferase*

1.1.2 *Protein and DNA adducts of heterocyclic aromatic amines.* The carcinogenic
N-hydroxy metabolites of HAAs or their *N*-sulfate and *N*-acetoxy esters can react and
bind to protein or DNA[18-35]. Protein-carcinogen adducts of several different classes or
carcinogens have been employed as biomarkers and successfully applied to monitor
human exposure[2]. Because of their abundance, long life-time, and ease of access non-
invasively, hemoglobin (Hb) and serum albumin (SA) are ideal proteins to employ as
biomarkers[2]. Assuming that the Hb- or SA-carcinogen adducts are stable, and that they
follow predictable kinetics, these protein adducts may provide an index of long-term
exposure to a contaminant over the life-time of the protein. However, it is necessary to
demonstrate the protein adduct is derived from the carcinogenic metabolite, and that
the adduct is a reflective measure of DNA adduct formation and toxicity in target
tissues. It is also essential that the domain of carcinogen adduct binding and the amino
acid sequence of the protein in animal models under investigation is the same as that in
humans.

The *N*-hydroxy metabolites of IQ and 2-amino-3,8-dimethylimidazo[4,5-
f]quinoxaline (MeIQx), following oxidation to the nitroso derivative, have been shown
to bind to SA to form a sulfinic acid amide linkage with a sulfhydryl-cysteine residue[18-
20,23]. The *N*-hydroxy-HAA metabolites may also react with oxyhemoglobin through a
co-oxidation process to form the arylnitroso metabolite and methemoglobin[2,18]. The
nitroso metabolites can then selectively bind to the β-93 sulfhydryl-cysteine residue
to form a sulfinamide linkage. These HAA-protein sulfinamide linkages appear to be
stable *in vivo*, but can be cleaved to regenerate the parent amine *in vitro* (**Scheme 1**).
Highly sensitive gas chromatography acid - mass spectroscopy methods have been
developed to measure for the acid cleaved amines[2,23]. However, based upon studies
with rodents, non-human primates and a pilot study in humans[18-25], it appears that the
amount of HAAs bound to blood proteins as sulfinamide adducts is too low to
successfully biomonitor human populations.

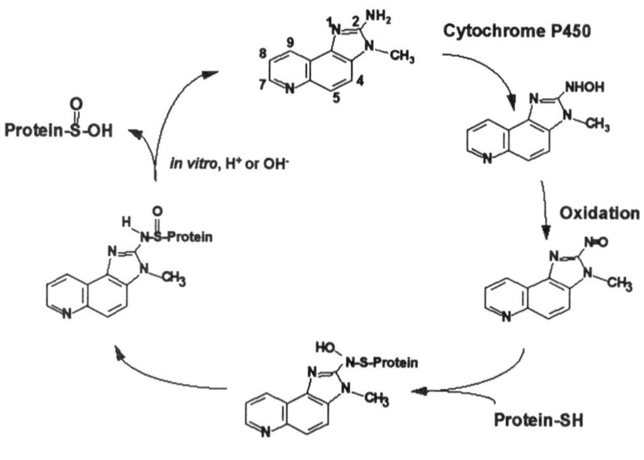

Scheme 1 *IQ-protein sulfinamide adduct formation in vitro*

The most direct evidence of recent exposure and genetic damage is through
measurement of DNA carcinogen adducts in cells[1,3-6]. In addition to monitoring DNA

carcinogen adducts *in vivo*, the eliminated products of DNA adducts may be measured in urine[6]. In the case of HAAs, DNA adduct formation has been shown to form principally with deoxyguanosine (dG) through the exocyclic amino groups of the HAAs to form dG-C8-HAA adducts[26-30]. A second adduct has been reported for IQ and MeIQx where it has been shown that adduction occurs at the C-5 group of the HAA and the 2-amino group of dG guanine to form dG-N^2-HAA adducts (Figure 3). The ^{32}P postlabelling method developed by Randerath and colleagues[31], which entails the enzymatic digestion of adducted DNA to 3'-deoxynucleotides followed by an enzymatic transfer of γ-^{32}P-ATP to the 5'-OH of the deoxyribose moiety, is perhaps the most widely used method to measure for DNA carcinogen adducts. Once labelled with γ-^{32}P-ATP, the DNA adducts are then resolved by thin layer chromatography or high performance liquid chromatography.

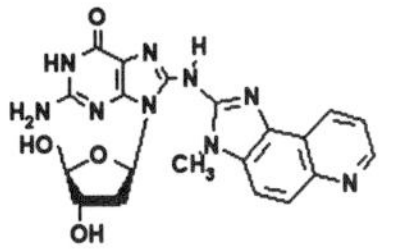

N-(deoxyguanosin-8-yl)-2-amino-3-methyl-
imidazo[4,5-*f*]quinoline (dG-C8-IQ)

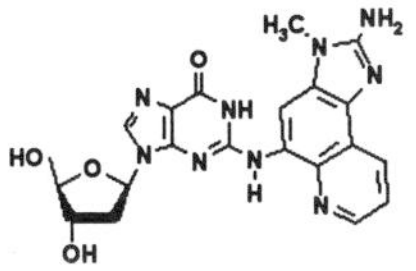

5-(deoxyguanosin-*N*²-yl)-2-amino-3-methyl-
imidazo[4,5-*f*]quinoline (dG-*N*²-IQ)

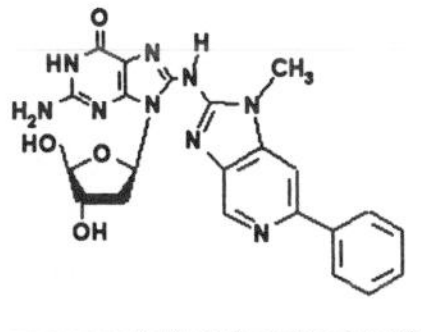

N-(deoxyguanosin-8-yl)-2-amino-1-methyl-
6-phenylimidazo[4,5-*b*]pyridine (dG-C8-PhIP)

Figure 3 *C-8 and N²-guanine adducts of IQ and C-8 guanine adduct of PhIP*

Numerous postlabelling studies on HAA-DNA adduct formation in animals have been reported[29]. The dG-C8 adducts are the predominant lesions followed by the dG-N^2 adducts. Recent data reveals that the the dG-N^2 adduct of IQ is more persistent than the dG-C8 adduct in liver and kidney of rats and it is this minor adduct which accumulates during chronic exposure to become the predominant lesion[32]. The dG-C8 adduct of 2-amino-1-methyl-6-phenylimidazo[4,5-*b*]pyridine (PhIP) has been detected in colo-rectal tissue of humans by the ^{32}P postlabelling method and by gas chromatography-mass spectroscopy[33]. HAA-DNA adduct formation in lymphocytes of rodents and non-human primates has been reported[34,35]; however, biomonitoring of HAA-DNA adducts in lymphocytes of humans has not been reported and merits investigation.

1.1.3 *Urinary metabolites of heterocyclic aromatic amines.* The major routes of metabolism and disposition of HAAs have been reported in animals[19,36]. Both urine and feces are major excretory routes of HAA metabolites. Metabolic pathways include cytochrome P450 mediated oxidation as well as phase II conjugation pathways. In addition to CYP catalyzed N-oxidation which leads to metabolic activation of HAAs, detoxification may occur through ring oxidation. Direct conjugation to the exocyclic amino group through glucuronidation and sulfamate formation also occurs and results in polar detoxification products. The pathways reported for MeIQx metabolism are shown in Figure 4.

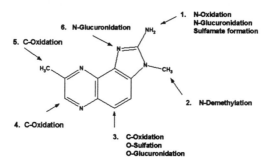

Figure 4 *Molecular sites of MeIQx metabolism*

Both MeIQx and PhIP have been identified in human urine following consumption of fried meat[37-41]. Acid hydrolysis of urine resulted in a 3 to 10-fold increase in the amount of MeIQx detected in urine and suggested the presence of acid labile sulfamate and N^1 or N^2 glucuronide conjugates of MeIQx[19,40,41](Figure 5). High performance liquid chromatography of these urinary metabolites showed the presence of both metabolites in human urine. The inter- and intra-individual variation in metabolism and formation of these conjugated metabolites may be in part attributed to enzyme polymorphisms or differences in enzyme induction due to dietary or environmental factors[38-43]. The contribution of these phase II detoxification pathways in humans appears similar to that reported in the rat given MeIQx at low doses and suggests that the rat is a good surrogate model to develop methods of human biomonitoring of HAAs.

2 CONCLUSION

The major routes of metabolism of HAAs have been identified and a number of protein and DNA adduction products have been characterized in animal models. Based upon metabolism studies *in vitro*, human tissues also are capable of activating HAAs at levels comparable to animal species which succumb to tumorigenesis during long-term feeding studies with HAAs. Therefore, exposure of HAAs through many daily staples is significant and these chemicals may be involved in the etiology of various human cancers.

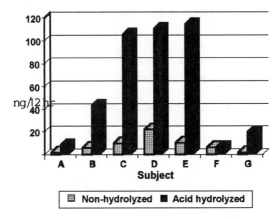

Figure 5 *MeIQx content in human urine following consumption of fried meat or fish*

Analysis of HAA urinary metabolites in humans following consumption of fried meat have shown that humans absorb HAAs. In the case of MeIQx, detoxification through glucuronidation and N^2-sulfamate formation occurs and demonstrates that rodents, non-human primates and humans detoxify this HAA through similar pathways. Measurement of HAAs and their phase II conjugates in urine provides a good index of exposure and absorption; however, these measurements do not provide information about metabolic activation and the biologically effective dose. Analytical methods are required to detect N-oxidation metabolites, such as the stable *N*-glucuronide conjugates of *N*-hydroxy-MeIQx or PhIP. Comparative analyses of urinary *N*-hydroxy-HAA conjugates, detoxification products, and DNA adducts in surrogate tissues, may provide useful biomarkers which will improve the extrapolation of animal toxicity data to humans, and the health risk posed by these dietary contaminants.

References

1. G.N. Wogan and N.J. Gorelick, *Environ. Health Perspect.*, 1985, **62**, 5.

2. P.L. Skipper and S.R. Tannenbaum, *Carcinogenesis*, 1990, **11**, 507.

3. D.E.G. Shuker and P.B. Farmer, *Chem. Res. Toxicol.*, 1992, **5**, 450.

4. R.J. Huggett, R.A. Kimberle, P.M. Mehrle Jr. and H.L. Bergman "Biomarkers. Biochemical, Physiological, and Histological Markers of Anthropogenic Stress" Lewis, Chelsea, MI, USA, 1992.

5. L.J. Marnett and P.C. Burcham, *Chem. Res. Toxicol.*, 1993, **6**, 771.

6. J.D. Groopman and T.W. Kensler *Chem. Res. Toxicol.*, 1993, **6**, 764.

7. N. Caporaso, M. T. Landi, and P. Vineis, *Pharmacogenetics*, **1**, 1991, 4.

8. K. Wakabayashi, M. Nagao, H. Esumi and T. Sugimura, *Cancer Res.* (Suppl), 1992, **52**, 2092s.

9. J.S. Felton and M.G. Knize "Handbook of Experimental Pharmacology" Springer-Verlag, Berlin and Heidelberg, 1990, Vol. 94/I, p. 471.

10. M. Jägerstad, K. Skog, S. Grivas, and K. Olsson, *Mutation Res.* 1991, **259**, 219.

11. G.A. Gross, R.J. Turesky, L.B. Fay, W.G. Stillwell, P.L. Skipper, and S.R. Tannenbaum, *Carcinogenesis*, 1993, **14**, 2313.

12. M.A. Butler, M. Iwasaki, F.P. Guengerich, and F.F. Kadlubar, *Proc. Natl. Acad. Sci, USA*, 1989, **86**, 7696.

13. Rich, K.J., Murray, B.P., Lewis, I., Rendell, N.B., Davies, D.S., Gooderham, N.J., and Boobis, A.R., *Carcinogenesis*, 1992, **13**, 2221.

14. R.J. Turesky, N.P. Lang, M.A. Butler, C.H. Teitel, and F.F. Kadlubar, *Carcinogenesis* 1991, **12**, 1839.

15. R. Kato, *CRC Crit. Rev. Toxicol.*, 1986, **16**, 307.

16. T. Sugimura, *Mutation Res.* 1988, **295**, 33.

17. R.H. Adamson, U.P. Thorgeirsson, E.G. Snyderwine, J. Reeves, D.W. Dalgard, S. Takayama, and T. Sugimura. *Jpn. J. Cancer Res.*, 1990, **81**, 10

18. R.J. Turesky, P.L. Skipper, and S.R. Tannenbaum, *Carcinogenesis* **8**, 1987, 1537.

19. R.J. Turesky, J. Markovic, I. Bracco-Hammer, and L.B. Fay, *Carcinogenesis*, 1991, **12**, 1847.

20. A.M. Lynch, S. Murray, A.R. Boobis, D.S. Davies, and N.J. Gooderham, *Carcinogenesis*, 1991, **12**, 1067.

21. A. Umemoto, Y. Monden, S. Grivas, K. Yamashita, and T. Sugimura, *Carcinogenesis*, 1992, **13**, 1025.

22. A. Umemoto, C. Negishi, S. Sato, and T. Sugimura, *Proc. Japan Acad. Ser. B*, 1986, **62**, 169.

23. A.M. Lynch, S. Murray, K. Zhao, N.J. Gooderham, A.R. Boobis, and D.S. Davies, *Carcinogenesis*, 1993, **14**, 191.

24. E.G. Synderwine, D.H. Welti, L.B. Fay, H.P. Würzner, and R.J. Turesky, *Chem. Res. Toxicol.* 1992, **5**, 843.

25. E.G. Snyderwine, D.H. Welti, C.D. Davis, L.B. Fay, and R.J. Turesky, *Carcinogenesis*, 1995, In Press.

26. R.J. Turesky and J. Markovic, *Chem. Res. Toxicol.* 1994, 7, 752.

27. D. Lin, K.R. Kaderlik, R.J. Turesky, D.W. Miller, J. O. Lay, Jr. and F.F. Kadlubar, *Chem. Res. Toxicol.* 1992, 5, 691.

28. H. Frandsen, S. Grivas, R. Andersson, L. Dragsted, and J.C. Larsen, *Carcinogenesis*, 1992, 13, 629.

29. R.J. Turesky, "DNA Adducts: Identification and Biological Significance", Lyon, International Agency for Research on Cancer, 1994, p. 217.

30. E.G. Snyderwine, P.P. Roller, R.H. Adamson, S. Sato, and S.S. Thorgeirsson, *Carcinogenesis*, 1988, 9, 1061

31. K. Randerath, E. Randerath, H..P. Agrawal, R.C. Gupta, M.E. Schurdak, and M.V. Reddy, *Environ. Health Perspect.*, 1985, 62, 57.

32. R. J. Turesky and J. Markovic, *Proceedings of American Assoc. Cancer Res.*, 1995, 36, 134.

33. M.D. Freissen, K. Kaderlik, D. Lin, L. Garren, H. Bartsch, N.P. Lang, and F.F. Kadlubar, *Chem. Res. Toxicol.* 1994, 7, 733.

34. D.A. Cummings, and H.A.J. Schut, *Carcinogenesis*, 1994, 15, 2623.

35. R.H. Adamson, E.G. Snyderwine, U.P. Thorgeirsson, H.A.J. Schut, R.J. Turesky, S.S. Thorgeirsson, S. Takayama, and T. Sugimura, "Xenobiotics and Cancer", Japan Scientific Society Press, Tokyo, 1991, p. 289.

36. J. Alexander, and H. Wallin, "Mutagens in Food: Detection and Prevention", Boca Raton. Fl., CRC Press, 1991, p. 143.

37. H. Ushiyama, K. Wakabayashi, M. Hirose, H. Itoh, T. Sugimura, and M. Nagao, *Carcinogenesis*, 12, 1991, 1417.

38. A.M. Lynch, M.G. Knize, A.R. Boobis, N.J. Gooderham, D.S. Davis and S. Murray, *Cancer Res.*, 1992, 52, 6216-6223.

39. A.R. Boobis, A.M. Lynch, S. Murray, R. De la Torre, A. Solans, M. Farré, J. Segura, N.H. Gooderham, and D.S. Davies, *Cancer Res.*, 1994, 54, 89.

40. W.G. Stillwell, R.J. Turesky, G.A. Gross, P.L. Skipper and S.R. Tannebaum *Cancer Epidemiology, Biomarkers & Prevention*, 1994, 3, 399.

41. R. J. Turesky, G.A. Gross, W.G. Stillwell, P.L. Skipper, and S.R. Tannenbaum, *Envron. Health Perspect.* 1994, 102, 47.

42. M.A. Butler, N.P. Lang, J.F. Young, N.E. Caporaso, P. Vineis, R.B. Hayes, C.H. Teitel, J.P. Massengill, M.F. Lawsen, and F.F. Kadlubar, *Pharmacogenetics*, **2**, 1992, 116.

43. R. Sinha, N. Rothman, E.D. Brown, S.D. Mark,, R.N. Hoover, N.E. Caporaso, O.A. Levander, M.G. Knize, N.P. Lang, and F.F. Kadlubar, *Cancer. Res.*, 1994, **54**, 6154.

Antioxidant Activity of Green Tea in Smokers and Non-smokers

B. J. P. Quartley, M. N. Clifford, R. Walker, and C. M. Williams

SCHOOL OF BIOLOGICAL SCIENCES, UNIVERSITY OF SURREY, GUILDFORD, SURREY GU2 5XH, UK

1 INTRODUCTION

There is now considerable interest in the role played by free radicals in many diseases[1] and the potential role of antioxidants in disease prevention. Dietary intake of naturally occurring compounds with antioxidant activity and their benefit to the consumer is therefore of increasing interest. Their potential benefit is suggested by epidemiological case control and cohort studies which have found dietary intake and plasma levels of the antioxidant vitamins (vitamins A, C, E and β carotene) to be negatively associated with development of cancer[2] and coronary heart disease.[3] However the effectiveness of dietary supplementation with antioxidant vitamins in chemoprevention appears, at present, unclear.[4] It must also be remembered that observational epidemiological studies are inherently vulnerable to confounding factors. The interest in naturally occurring compounds with antioxidant activity is, however, intense. One such group of compounds is the flavonoids. These are polyphenolic, non-nutrient, compounds of diverse subgroup which are ubiquitous in fruits and vegetables and present in some beverages. *In vitro* flavonoids have been shown to scavenge free radicals,[5,6] quench singlet oxygen,[7] prevent fatty acid autooxidation[8] and inhibit free radical-producing enzymes,[9,10] all of which could contribute to an overall antioxidant effect.

Green tea is a commonly consumed beverage in some areas and is of particular interest. It is rich in a group of flavonoids, the flavan-3-ols, termed catechins, present at up to 30% dry weight. *In vitro,* green tea extracts have been shown to possess radical scavenging activity[11] and to protect fatty acids from autooxidation.[12] In animal studies they have been shown to protect against chemically induced oxidative damage,[13,14] spare vitamin E[15] and increase the activity of antioxidant enzymes.[16] Similar actions have also been observed in humans with hepatitis.[17] These studies have, however, involved either high doses of green tea or severe forms of free radical exposure. Our aim was therefore to investigate the antioxidant activity of a moderate, and in dietary terms realistic, dose of green tea in ostensibly healthy free living human volunteers.

The study was carried out in both smokers and non-smokers given fish oils. Smokers are exposed to free radicals from cigarette smoke and tar[18] and from the proinflammatory actions of smoking.[19,20] Smokers also have lower plasma concentrations of vitamin C[21] and β carotene,[22] due to decreased dietary intake[23] and possibly increased turnover *in vivo.*[24] The endogenous antioxidant defence of smokers might therefore be compromised putting these individuals at an increased risk of oxidative damage resulting from free radical exposure. The polyunsaturated fatty acids. eicosapentaenoic acid (EPA) and docosahexaenoic acid (DHA), in fish oils are more susceptible to hydrogen abstraction and lipid peroxidation than more saturated fatty acids. Fish oils are therefore more susceptible to oxidation from free radicals and active oxygen species produced from basal metabolism *in vivo*. Smoking and fish oil supplementation are thus two situations which might tax the body's endogenous antioxidant defences, especially in combination, leading to oxidative stress and increased levels of products of oxidative attack on proteins, DNA and lipids. Indeed smoking has been shown to increase lipid peroxidation[25] and DNA oxidation,[26,27] while it is well documented that fish oil supplementation increases lipid peroxidation.[28,29,30] These are situations in which additional antioxidants might be required and in which green tea has a greater opportunity to demonstrate *in vivo* antioxidant activity.

2 METHODS AND MATERIALS

2.1 Subjects

Thirty men aged 40-60 from the Guildford area volunteered for the study and were screened by an initial interview and questionnaire. Volunteers taking any form of vitamin or mineral supplement, with a history of heart disease, cancer or any chronic illness, drinking more than 21 units alcohol/week, eating oily fish more than 3 times per week, eating more than 4 portions of fruit and vegetables/day, taking more than three 30 minute periods of vigorous physical exercise per week or consuming green tea or any form of herbal preparation were excluded. From this group 9 non-smokers and 7 smokers were accepted onto the study.

2.2 Study design and treatment

The study was a placebo-controlled single blind cross over design. Subjects received 10 g of the fish oil MaxEPA (19·5% EPA, 10.2% DHA, 1 mg/g vitamin E, Sevenseas Ltd, Hull) for two periods of 4 weeks with a 3 week wash out between. During one period they also received 1·5 g green tea in capsular form while during the other they received outwardly identical capsules containing 1·5 g of placebo (maltodextrin). The green tea used was a freeze dried extract containing approximately 30% catechins. Subjects took three 1 g fish oil capsules and two 250 mg capsules of green tea or placebo after each main meal (four fish oil capsules after breakfast). Subjects were instructed to maintain their normal diet throughout the study with the exception that red wine and black tea were prohibited from 1 week prior to the study and throughout the 11 week study period.

2.3 Sample collection

To monitor the effects of the treatment samples of blood, breath and urine were collected prior to each treatment period, after 2 weeks and at the end of each period (4 weeks) for the measurement of TBARS in urine and plasma and exhaled ethane in breath. Blood and breath samples were collected in the morning after an overnight fast and for smokers after at least 8·5 hours abstention from smoking. Subjects were instructed to refrain from vigorous physical exercise or alcohol consumption on the day prior to sample collection. During the study, subjects did not take their breakfast dose of capsules until samples had been collected. Breath samples were collected into cali-5-bond 5 layer gas sampling bags (Alltech Associates Ltd, Lancashire). Subjects breathed Synthetic Air (<0·1 ppm total hydrocarbons) (BOC Ltd) for 4 minutes prior to sample collection to remove any atmospheric air from the lungs. The BOC synthetic air used was found to contain no detectable ethane and was supplied from Douglas bags (Harvard Apparatus Ltd, Kent); these were found to be free from ethane contamination at least 2 hours after filling. Breath samples in gas sampling bags remained free from ethane contamination at least 24 hours after filling. Blood samples were collected into EDTA tubes (Sarstedt Ltd, Leicester). Twenty-four hour urine samples were collected into 1 M hydrochloric acid on the day prior to sample collection.

2.4 Ethane analysis

Ethane analysis was performed using a Perkin Elmer 8500 gas chromatograph equipped with a Valco gas sampling valve and 165 ml sampling loop and flame ionisation detector. Separation was performed using a 6' × 1/8" Carbosphere 80/100 column (Alltech Associates Ltd, Lancashire) with nitrogen carrier gas at a flow rate of 35 ml/min. All gases were scrubbed by passage through activated charcoal traps. The sample loop was filled with breath samples and the loop contents flushed onto the column at 40 °C. The column was maintained at this temperature for 6·5 minutes while the sample was applied. To elute the ethane peak a temperature gradient was employed. The column was heated to 220 °C at 30 °C/min and held at this temperature for 5 minutes. Ethane eluted at a retention time of 16·15 minutes. The column was then heated to 330 °C and held at this temperature for 3 minutes. This was found necessary to ensure good peak shapes in subsequent runs.

2.5 Thiobarbituric acid reactive substances (TBARS) assay of plasma and urine

Analysis was performed on plasma samples within 1 hour of blood collection and on urine samples after centrifugation at 3000 rpm for 10 minutes using essentially the same procedure. To 0·5 ml of plasma or urine sample 0·1 ml of 0·73 M butylated hydroxytoluene (BHT) in ethanol and 5 ml saturated diethylthiobarbituric acid (DETBA) in 0·1 M phosphate buffer (pH 3·25) were added. After boiling for 1 hour samples were allowed to cool for 30 minutes. Butanol (3 ml) was added and samples vortexed for 10 seconds. After centrifugation for 5 minutes at 2500 rpm the butanol layer was removed and the process repeated with a further 2 ml and the butanol extracts pooled. Extracts could be stored for at least 3 weeks in the dark at 4 °C. Prior to HPLC analysis 0·5 ml butanol extracts were evaporated under nitrogen at 37 °C and the residue resuspended in acetonitrile/water (50:50). Separation was carried out on a 25 cm Kromasil C_{18} column (Hichrom Ltd, Berkshire) connected to a Spectra Physics XR pump and AS 100 autosampler. The diethylthiobarbituric acid-malondialdehyde ($DETBA_2$-MDA) adduct was eluted isocratically using 50% acetonitrile and 50% water containing 0·1% triethanolamine as solvent. Detection was at 540 nm using Spectraphysics UV 2000 detector. The identity and purity of the $DETBA_2$-MDA peak from plasma and urine samples was confirmed by comparison of retention time and spectral data with those obtained from 1,1,3,3-tetraethoxypropane standard.

3 RESULTS

Both periods of treatment were well tolerated by the smokers and non-smokers. Subject characteristics are shown in Table 1. There was no significant difference in age, weight, BMI or alcohol consumption between the two groups. The smokers had all smoked for at least 10 years prior to the study while the non-smokers consisted of both ex-smokers and those who had never smoked. The ex-smokers had given up at least 8 years prior to the study.

Table 1 *Subject characteristics*

	Non-smokers (n=9)	*Smokers (n=7)*
Age (years)	51·1±6·8	45·6±6·7
Weight (kg)	83·7±11·7	82·4±18·2
BMI (kg/m²)	27·3±2·5	26·1±4·5
Alcohol		
units/week	6·5±6·0	3·4±3·7
Cigarettes		
number/day	–	23·6±9·9
years smoked	–	18·9±8·0

Fish oil supplementation with green tea or placebo had no statistically significant effect on ethane exhalation in either smokers or non-smokers (Table 2). Similarly there was no statistically significant difference between the periods of fish oil supplementation with green tea and placebo in either group. Ethane exhalation was however consistently greater in the smokers compared to the non-smokers throughout the study.

Table 2 *Effect of green tea on ethane exhalation in smokers and non-smokers given fish oils*

	pmol ethane/l breath			
	Non-smokers		*Smokers*	
	Fish oil-Tea	Fish oil-Placebo	Fish oil-Tea	Fish oil-Placebo
Baseline	48±34	42±17a	92±50	120±64
2 weeks	39±13a	57±27	78±33	86±46
4 weeks	32±18b	34±14a	94±66	91±48

Letters indicate significant difference from corresponding group of smokers (a *P<0·01*, b *P<0·02*)

Fish oil supplementation had no statistically significant effect on plasma TBARS in the non-smokers and there was no significant difference between the green tea and placebo periods in this group (Table 3). In the smokers, plasma TBARS increased during fish oil supplementation with both placebo and green tea. This effect reached statistical significance during the placebo period only, but there was no significant difference between green tea and placebo periods. The effect of fish oil supplementation on urine TBARS is shown in Table 4. After 2 weeks, urine TBARS were significantly greater than baseline in smokers during the green tea and placebo periods and in the non-smokers during the green tea period only. After 4 weeks of supplementation there was a decrease in urine TBARS and values were now no longer statistically significantly different from baseline. The increase in urine TBARS was consistently greater in the smokers compared to the non-smokers. Green tea had no significant affect on urine TBARS in either smokers or non-smokers.

Table 3 *Effect of green tea on plasma TBARS in smokers and non-smokers given fish oil*

| | nmol malondialdehyde/ml plasma | | | |
| | Non-smokers | | Smokers | |
	Fish oil-Tea	Fish oil-Placebo	Fish oil-Tea	Fish oil-Placebo
Baseline	0·67±0·14	0·67±0·18	0·54±0·27	0·51±0·09
2 weeks	0·69±0·18	0·77±0·11	0·83±0·35	0·90±0·20*
4 weeks	0·72±0·17	0·70±0·16	0·60±0·20	0·72±0·20**

Asterisk indicates significant difference from baseline (* $P<0·01$, ** $P<0·05$)

Table 4 *Effect of green tea on urine TBARS in smokers and non-smokers given fish oil*

| | µg malondialdehyde/kg/24 hours | | | |
| | Non-smokers | | Smokers | |
	Fish oil-Tea	Fish oil-Placebo	Fish oil-Tea	Fish oil-Placebo
Baseline	5·38±1·41	5·98±1·70	6·19±2·52	5·90±1·35
2 weeks	7·35±2·26*	7·05±2·18a	9·23±3·71**	10·45±2·66***
4 weeks	7·15±1·86	6·37±1·87	7·15±2·75	8·02±2·35

Letters indicate significant difference from corresponding group of smokers. (a $P<0·02$)
Asterisk indicates significant difference from baseline(* $P<0·02$, ** $P<0·05$, *** $P<0·001$)

4 DISCUSSION

The results of this study indicate that smokers consistently exhale greater quantities of ethane than non-smokers which is consistent with previous observations that smoking increases ethane exhalation.[31,32] In this study, mean values for ethane exhalation were upto three times greater in the smokers but showed substantial inter-individual variability which in some cases prevented the differences in mean values becoming statistically significant. With respect to smoking, non-smokers have a consistent behaviour whereas smokers differ in their frequency of indulgence and/or the nature of the tobacco consumed and thus can be expected to be more variable with respect to relevant biochemical markers.

Plasma and urine TBARS were however not significantly different in the two groups at baseline. A possible explanation for this lies in the nature of free radical exposure resulting from smoking. The lungs are the primary site of exposure to free radicals from cigarette smoke and smokers' lungs also accumulate large numbers of activated phagocytic cells.[33,34] Thus lipid peroxidation could occur to a much greater extent in smokers' lungs compared to the rest of the body. The greater ethane exhalation seen in smokers could therefore be coming mainly from lung cell lipid peroxidation rather than "whole body" oxidative stress. If this is the case other indices of lipid peroxidation measured in bodily fluids might be much less affected by smoking, explaining why at baseline, plasma and urine TBARS were not elevated in smokers compared to non-smokers.

The increase in plasma and urine TBARS without any corresponding increase in ethane exhalation on giving fish oils to smokers appears on first consideration to be inconsistent. A possible explanation for this may lie in the kinetics of lipid peroxidation resulting from fish oil supplementation which might be considered a two component model. The response to fish oils immediately after ingestion might represent an acute phase while incorporation of long chain fatty acids into cellular membranes, displacing more saturated fatty acids[35] and rendering the membrane more susceptible to lipid peroxidation, might represent a chronic phase. It is possible that the increased lipid peroxidation on giving fish oils is primarily due to the acute phase. Not only is the body presented with more peroxidisable substrate susceptible to oxidation by cyclooxygenase and lipoxygenase enzymes and free radicals produced from basal metabolism, but the free radicals produced could also initiate lipid peroxidation of cellular membranes, especially if they contain increased levels of EPA and DHA. However, once the dose of fish oil has been metabolised there will be less peroxidisable substrate immediately available and less free radicals produced from this source to initiate further membrane lipid peroxidation, both of which might result in a reduction in lipid peroxidation compared to the acute phase. The overall effect of fish oil supplementation might be an increased background level of lipid peroxidation associated with EPA and DHA deposition in cellular membranes accompanied by a series of acute peaks of lipid peroxidation associated with the immediate metabolism of each dose of fish oil. Urine samples were collected over a 24 hour period during which the subjects took 10 g of MaxEPA and therefore include the acute phase. In contrast, plasma and breath samples were collected approximately 12 hours after the subjects took their last dose of fish oil and might be considered to represent the chronic phase only. The increase in plasma TBARS without any increase in ethane exhalation is not obviously consistent with this explanation. It must be remembered, however, that the exhaled ethane is a single terminal product of lipid peroxidation and assuming that there are no significant fluctuations in factors affecting the tissues accumulation and/or release of ethane, the quantity detected represents oxidative events that have reached completion at or around the time of sampling. In contrast the TBARS assay measures lipid hydroperoxides, which decompose during the assay, as well as preformed malondialdehyde. The production of malondialdehyde is thus greatly amplified while ethane exhalation, representing only a minor end product of lipid peroxidation, is not. It is also possible that any non-peroxidised EPA and DHA present in plasma samples could be peroxidised during the TBARS assay itself. This is one of the limitations of the TBARS assay and would increase plasma TBARS without affecting ethane exhalation. The lipid soluble antioxidant butylated hydroxytoluene was, however, added to plasma samples in accordance with the recommendations of Hoving *et al.* (1992)[36] to minimise this.

Mean plasma and urine TBARS in the smokers given fish oil with both placebo and green tea were consistently lower after 4 weeks of supplementation compared to the 2 week values. Similarly, Brown *et al.* (1990)[29] found 15 g MaxEPA/day significantly increased plasma TBARS after 2 weeks of supplementation in non-smokers but values were not significantly different from baseline after 4 weeks. In a more long term study Meydani *et al.* (1991)[37] found a recovery in plasma TBARS between 2 and 3 months of fish oil supplementation. Induction of antioxidant defence mechanisms is one possible explanation for this, as has been previously suggested. Indeed Bellisola *et al.* (1992)[38] found a statistically significant increase in glutathione peroxidase activity after 10 weeks of fish oil supplementation.

The effect of fish oil supplementation on lipid peroxidation in the non-smokers in this study was generally small and the only significant increase seen was in urine TBARS after 2 weeks of fish oil supplementation with green tea. This value subsequently declined and was not significantly different from baseline after 4 weeks. The endogenous antioxidant defence of the non-smokers therefore appears almost able to deal with this additional oxidant load and, in the absence of obvious oxidative stress, it is not surprising that the relatively low dose of green tea given had no measurable antioxidant activity. Fish oil supplementation did, however, cause a greater increase in plasma and urine TBARS in the smokers. The antioxidant defence of smokers therefore appears inadequate to protect against the increased lipid peroxidation in this situation and smokers appear to require additional antioxidants. It is under conditions such as these that green tea is more likely to demonstrate antioxidant activity. Indeed mean values for both plasma and urine TBARS in smokers were lower during the fish oil and green tea period compared to the corresponding placebo period. This effect was seen after both 2 and 4 weeks supplementation in both TBARS assays but never reached statistical significance. The dose of green tea given was small and will only have had a finite antioxidant capacity, especially if its principal mode of action is through direct radical scavenging or regeneration of vitamins C or E. It may be speculated that if a higher dose of green tea had been given the effect seen might have been greater. Similarly, for the non-smokers a larger dose of fish oil might have compromised their antioxidant defence giving more potential for the green tea to exert demonstratable antioxidant activity. The aim of this study was, however, to investigate the antioxidant activity of a realistic dose of green tea.

In conclusion smoking appears to compromise the body's antioxidant defence putting smokers at increased risk from oxidative damage caused by free radicals. As measured by the biomarkers used in this study, green tea had no significant antioxidant activity against lipid peroxidation. This does not rule out the possibility that other indices of oxidative damage might be reduced by green tea. Analyses are currently underway investigating oxidative damage to DNA and proteins, and plasma concentrations of plasma vitamins C, E and β carotene.

References

1. J. M. C. Gutteridge, *Free. Rad. Res. Comms.*, 1993, **19**, 141.
2. M. L. Burr, *J. Hum. Nut & Diet.*, 1994, **7**, 409.
3. J. M. Gaziano, *Am. J. Med.*, 1994, **97**, 3A.
4. The alpha-tocopherol, beta carotene cancer prevention study group. *N. Engl. J. Med.*, 1994, **330**, 1029.
5. J. Robak and R. J. Gryglewski, *Biochem. Pharmacol.*, 1988, **37**, 837.
6. S. F. Husain, J. Cillard and P. Cillard, *Phytochem.*, 1987, **26**, 2489.
7. C. Tournaire, S. Croux and M. Maurette, *J. Phytochem. Photobiol. B: Biol.*, 1993, **19**, 205.
8. Y. Fujita, K. Komagoe, Y. Niwa, I. Uehara, R. Hara, H. Mori, T. Okuda and T. Yoshida, *Yakugaku Zasshi.*, 1988, **108**, 528.
9. J. Pincemail, C. Deby, A. Thirion, M. de Bruyn-Dister and R. Goutier, *Experimentia.*, 1988, **44**, 450.
10. U. Takahama, *Phytochem.*, 1985, **24**, 1443.
11. B. Zhao, X. Li, R. He, S. Cheng and X. Wenjuan, *Cell. Biophys.*, 1989, **14**, 175.

12. C-T. Ho, Q. Chen, H. Shi, K-Q. Zhan and R. T. Rosen, *Prev. Med.*, 1992, **21**, 520.
13. C. G. Fraga, V. S. Martino, G. E. Ferraro, J. D. Coussio and A. Bovaris, *Biochem. Pharmacol.*, 1987, **36**, 717.
14. I. Edes, A. Toszegi, M. Csanady and B. Bozoky, *Cardiovas. Res.*, 1986, **20**, 542.
15. F. Nanjo, M. Honda, K. Okushio, N. Matsumoto, F. Ishigaki, T. Ishigami and Y. Hara, *Biol. Pharm. Bull.*, 1993, **16**, 1156.
16. S. G. Khan, S. K. Katiyar, R. Agarwal and H. Muktar, *Cancer. Res.*, 1992, **52**, 4050.
17. A. Par, M. Mezes, P. Nemeth and T. Javor. *Int. J. Clin. Pharm. Res.*, 1985, **6**, 389.
18. D. F. Church and W. A. Pryor, *Environ. Health. Perspec.*, 1985, **64**, 111.
19. M. C. Yeung and A. D. Buncio, *Am. J. Med.*, 1984, **76**, 31.
20. J. R. Hoidal and D. E. Niewoehner, *Am. Rev. Respir. Dis.*, 1982, **126**, 548.
21. G. Schectman, J. C. Byrd and R Hoffmann, *Am. J. Clin. Nutr.*, 1991, **53**, 1466.
22. W. S. Stryker, L. A. Kaplin, E. A. Stein, M. J. Stampfer, A. Sober and W. C. Willett, *Am. J. Epidemiol.*, 1988, **127**, 283.
23. C. Bolton-Smith, *Ann. N. Y. Acad Sci.*, 1993, **686**, 347.
24. A. B. Kallner, D. Hartmann, D. H. Homig, *Am. J. Clin. Nutr.*, 1981, **34**. 1347.
25. E. Hoshino, R. Shariff, A. Van Gossum, J. P. Allard, C. Pichard, R. Kurian and K. N. Jeejeebhoy, *J. P. E. N.*, 1990, **14**, 300.
26. S. Loft, K. Vistisen, M. Ewertz, A. Tjonneland, K. Overvad and H. E. Poulsen, *Carcinogenesis.*, 1992, **13**, 2241.
27. H. Kiyosawa, M. Suko, H. Okudaira, K. Murata, T. Miyamoto, M-H Chung, H. Kasai and S. Nishimura, *Free. Rad. Res. Comms.*, 1990, **11**, 23.
28. D. Harats, Y. Dabach, G. Hollander, M. Ben-Naim, R. Schwartz, E. M. Berry, O. Stein and Y Stein, *Atherosclerosis*, 1991, **90**, 127.
29. J. E. Brown and K. W. J. Whale, *Clinica, Chimica, Acta.*, 1990, **193**, 147.
30. O. Haglund, R. Luostarinen, R. Wallin, L. Wibell and T. Saldeen, *J. Nutr.*, 1991, **121**, 165.
31. C. Wade, A. M. van Rij, *Proc. Univ. Otago. Med. Sch.*, 1986, **64**, 75.
32. M. Sakamoto, *Nippon. Eisseigaku. Zasshi.*, 1985, **40**, 835.
33. R. C. Hubbard, F. Ogushi, G. A. Fells, A. M. Cantin, S. Jallet, M. Courtney and R. G. Crystal, *J. Clin. Invest.*, 1987, **80**, 1289.
34. J. R. Hoidal, R. B. Fox, P. A. LeMarbe, R. Perri and J. E. Repine, *Am. Rev. Resp. Dis.,* 1981, **123,** 85.
35. C. E. Hock, M. A. Holahan and D. K. Reibel, *Am J Physiol.*, 1987, **252**, H554-H560.
36. E. B. Hoving, C. Laing, H. E. Rugers, M. Teggler, J. J. van Doormaal and A. J. Muskiet, *Clinica. Chimica. Acta.*, 1992, **208**, 63.
37. M. Maydani, F. Natiello, B. Goldin, N. Free, M. Woods, E. Schaefer, J. B. Blumberg and S. L. Gorbach, *J. Nutr.*, 1991, **121**, 484.
38. G. Bellisola, S. Galassini, G. Moschini, G. Poli, G. Perona and G. Guidi, *Clinica. Chimica. Acta.*, 1992, **205**, 75.

Non-invasive or Minimally Invasive Biomarkers of Exposure to Genotoxic Agents Derived from Food

D. E. G. Shuker, C. Leuratti, K. Harrison, J. Conduah Birt, and P. B. Farmer

MRC TOXICOLOGY UNIT, HODGKIN BUILDING, UNIVERSITY OF LEICESTER, PO BOX 138, LANCASTER ROAD, LEICESTER LE1 9HN, UK

1 INTRODUCTION

The incidence of some of the major cancers (of gastrointestinal tract [such as esophagus, stomach and colon/rectum] and breast) varies widely over the world[1]. The sometimes extreme extent of the variation (for example, of esophageal cancer within China where there is an 20-fold difference between the highest and lowest rates) has led to the conclusion that environmental and/or lifestyle factors have a major role in their etiology[2]. Striking evidence for this conclusion has come from studies of migrant populations, such as Japanese living in Hawaii or on the West Coast of the United States, where their cancer rates, at sites such as stomach, begin to reflect those of the local population within one generation[3]. There is good evidence that changes in consumption of specific dietary components are linked to these effects[4]. The underlying etiology of the major cancers noted above has not been unambiguously elucidated but many epidemiological studies have highlighted various dietary risk, and protective factors which account for some of the variability. High consumption of fat and red meat are risk factors for colorectal cancers and a lack of fresh fruits and vegetables is associated with increasing risk of stomach cancer.

Many of the known human carcinogens are DNA-damaging agents and there is increasing evidence that the formation of covalent DNA adducts is a necessary, but probably not suffcient, factor in the mechanisms of carcinogenesis of these agents[5]. Given that the human diet is an extremely complex mixture of chemicals it is extremely unlikely that the risk of cancer at any site will be associated with single food constituents. However, this can sometimes happen in the case of contamination of food with particularly potent carcinogens, such as the naturally occuring aflatoxins, which are responsible for otherwise rare cancers. It may be the effect of diet on the levels of endogenous DNA damage such as depurination, oxidation and deamination which accounts for the major cancer risks associated with diet[6]. Conversely, the modulation of these same endogenous processes by dietary factors may account for the protection against cancer provided by some foods such as fresh fruits and vegetables[7]. The unique aspects of dealing with food-related exposures to genotoxic agents has been succinctly stated by Saracci[8], "Unlike xenobiotic agents which, if found to be carcinogenic, can at least in principle be dispensed with, food is indispensable and many individual nutrients cannot be dispensed with either. Furthermore, in contrast with a toxic xenobiotic agent for which the lowest possible exposure is best for health, an optimal range of intake exists for most nutrients, and this needs to be determined as accurately as possible in humans so that the levels above and below which harmful effects occur can be identified".

We have recently begun a programme of work aimed at developing a number of biomarkers of exposure to food-related DNA-damaging agents. The overall objectives of the programme are to develop biomarkers of DNA damage derived from macroscopic components of diet (fat and protein) as well as certain individual food components consumed at a relatively high level, such as alcohol. The approaches that are being employed are based on the use of accessible bodily fluids (blood and urine) and the analysis of covalently-modified macromolecules such as haemoglobin and excreted DNA adducts. These approaches have been reviewed in a number of articles[9,10]. The analytical techniques which have been most widely used in our laboratory have also been reviewed and include mass spectrometry[11] and immunoaffinity purification in combination with various analytical methods[12].

2   BIOMARKERS OF EXPOSURE TO NITROSATION PRODUCTS OF AMINO ACIDS AND PEPTIDES

Amongst the many N-alkyl-N-nitroso compounds that are known to be mutagenic and carcinogenic, there are a number which share a common feature of being derived from the simplest amino acid glycine and are all, in principle, carboxymethylating agents[13]. N-nitrosoglycocholic acid (NOGC) is a carcinogenic and mutagenic derivative of the naturally occuring bile acid conjugate, glycocholic acid[14,15,16]. Incubation of NOGC with calf thymus DNA in vitro gave rise to N-7-carboxymethylguanine (7-CMG), N-3-carboxymethyladenine   (3-CMA)   and O^6-carboxymethylguanine   (O^6-CMG)[17]. Furthermore, administration of [14C]-NOGC to rats gave rise to dose dependent excretion of 7-CMG in urine[17]. N-nitrosopeptides which are C-terminal in glycine, such as N-(N'-acetyl-L-prolyl)-N-nitrosoglycine (APNG), are mutagenic[18] and carcinogenic[19] and would be expected to be carboxymethylating agents by analogy with NOGC. Similarly, N-nitroso-N-carboxymethylurea is a gastrointestinal carcinogen[20,21]. Azaserine, a pancreatic carcinogen, is also known to carboxymethylate DNA *in vivo* and [14C]-7-CMG was detected in DNA extracted from pancreatic acinar cells treated with [14C]-azaserine[22].

Recently, O^6-carboxymethyl-2'-deoxyguanosine (O^6-CMdG) has attracted our particular interest because of its apparent lack of repair by bacterial and mammalian O^6-alkylguanine alkyl transferases (Shuker and Margison, unpublished results) and therefore offered the possibility of a group-specific and possibly persistent DNA adduct derived from nitrosated glycine derivatives. A specific antiserum to O^6-CMdG was prepared and used to make reusable immunoaffinity columns which selectively retained the adduct from DNA hydrolysates. The binding of O^6-CMdG was so strong that conditions used to elute the adduct (0.1 M trifluoroacetic acid) resulted in partial hydrolysis (becoming quantitative on heating) of the glycosidic bond to give O^6-CMG which is more fluorescent than the deoxynucleoside. DNA treated with APNG (5 mmol) afforded O^6-CMdG at levels of 35 pmol/mg DNA. Similar results were obtained with potassium diazoacetate (Harrison *et al.*, manuscript in preparation). The antiserum was also found to be capable of detecting O^6-CMdG *in situ* in tissue sections from azaserine-treated animals by immunohistochemical staining (Harrison *et al*, unpublished results).

In summary, a diverse range of N-nitrosoglycine or nitrosated glycine derivatives all give rise to the same DNA adduct, O^6-CMdG, *via* formation of a common carboxymethylating intermediate (Scheme 1). O^6-CMdG and related adducts (7-CMG and 3-CMA) are all potential markers of exposure to diet-related nitrosation pathways

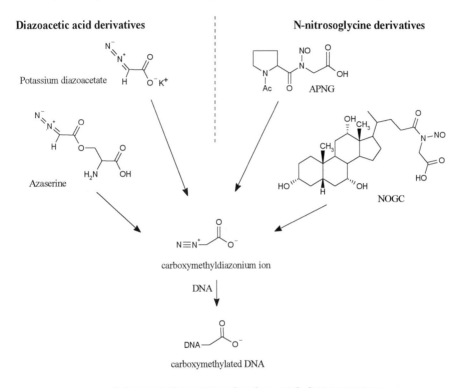

Scheme 1. *Formation of carboxymethylating agents from a range of nitrosated glycine derivatives*

3   MALONDIALDEHYDE-DNA ADDUCTS AS BIOMARKERS OF LIPID PEROXIDATION

Malondialdehyde (MDA) is the most abundant carbonyl compound and the major mutagenic and carcinogenic product generated by lipid peroxidation. The major DNA adduct formed at neutral pH is the highly fluorescent pyrimidopurinone product from 2'-deoxyguanosine (M_1dG, Scheme 2)[23,24]. Recent evidence suggests that M_1dG is present in human liver DNA at levels of 5-11 adducts per 10^7 bases[25]. The analytical method involved the isolation of the base M_1G from DNA hydrolysates followed by conversion to a pentafluorobenzyl derivative which was quantified by NICI GC-MS[26]. The limit of sensitivity of the assay was approximately 2 adducts per 10^8 base pairs with 300 µg of DNA. Previous work had suggested that both rats and humans excreted M_1G base in urine[26,27] but this could not be satisfactorily reproduced in another laboratory[28].

M_1dGMP has been detected in DNA using ^{32}P-postlabelling but the currently available methods have some drawbacks. The procedure developed by Vaca *et al*[29] does not, in our hands, satisfactorily separate M_1dGMP from the normal base dGMP. The recently published procedure of Wang *et al*[30] would appear to more suitable but still lacks the power to unambiguously quantify M_1dG.

We have improved the synthesis of M_1dGMP for use as an authentic standard and developed conditions for purification of the adduct by HPLC with fluorescence detection prior to postlabelling. M_1dGMP is phosphorylated by incubation with   [γ-^{32}P]-ATP and T4 polynucleotide kinase and the resulting 3',5'-bisphosphate separated by two-dimensional TLC. Based on our previous experience[12] we are also attempting to prepare antibodies to M_1dG in order to make immunoaffinity columns which will allow selective purification of the adduct.

dG M_1dG

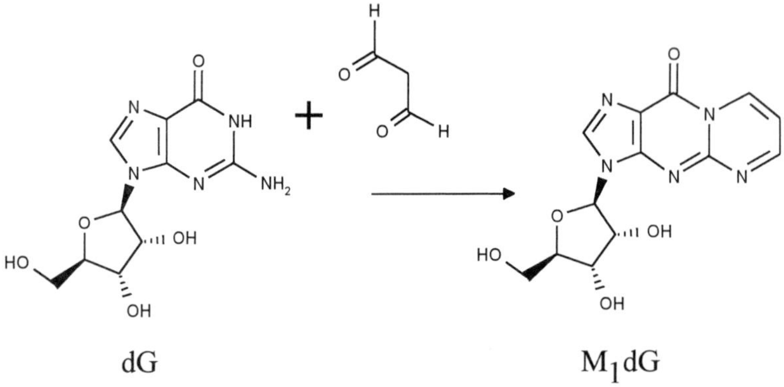

Scheme 2. *Reaction of malondialdehyde with deoxyguanosine*

4 STABLE ACETALDEHYDE-PROTEIN ADDUCTS AS BIOMARKERS OF ALCOHOL EXPOSURE

Acetaldehyde is the highly reactive intermediate produced in the metabolism of ethanol to acetate mainly by the action of alcohol dehydrogenase and microsomal cytochrome P450IIE1. The toxic effects of excessive alcohol consumption including liver disease are thought to be due to acetaldehyde, which binds to hepatic as well as other proteins. A number of *in vitro* studies have shown that acetaldehyde forms stable and unstable adducts with haemoglobin and proteins, such as albumin, tubulin and collagen[31]. In the case of haemoglobin, several types of adducts have been identified as due to acetaldehyde modification.

Various approaches such as hplc, nmr, ms and immunological assays have been used for detecting acetaldehyde adducts. Modified haemoglobin has been detected in red blood cells and blood from alcohol consuming individuals[31,32,33]. NMR and MS techniques have been utilized to confirm the formation and modification of haemoglobin and peptides[34,35,36]. In immunological studies, antibodies have been produced that recognize acetaldehyde modified proteins but the structures of the epitopes have not been characterised. However, at present there is no reliable quantitative method available for measuring acetaldehyde adducts. The aim of this study was to locate a site on the haemoglobin chain where a reproducible adduct is formed, identify the structure and develop a suitable quantitative method for analysis.

Preliminary studies were carried out using two model peptides, a dipeptide (valine-leucine), and a polypeptide (21 chain amino acid). Incubation of the two peptides with acetaldehyde yielded stable adducts which we were able to identify by mass spectrometry. Possible unstable adducts may have been produced as intermediates but these were not detected in these assays. From the tandem MS/MS analysis it appeared that the stable adducts were formed between the aldehyde and the amino group of the N-terminal valine of each of the peptides. San George and Hoberman[34] found that on incubation of haemoglobin with acetaldehyde an imidazolidinone type adduct was produced at the N-terminal end of haemoglobin and we have confirmed these results with the model peptides and shown by NMR that a diastereoisomeric mixture of adducts is formed (Scheme 3). Current work is aimed at developing a quantitative method for the analysis of the stable acetaldehyde adduct on the N-terminal valine of haemoglobin.

5 DISCUSSION

The work described in this short review is at an early stage of development but the preliminary results suggest that informative biomarkers of human exposure to genotoxic compounds, particularly alkylating agents derived from major constituents in diet, can be developed. The work of Ames and his collaborators[6] on DNA oxidation suggests that this endogenous source of genotoxic damage can be strongly influenced by diet, particularly total caloric intake. However, the focus of our work is on alkylation damage from endogenous sources which occurs at levels not dissimilar to those observed for oxidative DNA damage. Thus, macroscopic components of the diet such as fat and protein can be converted by endogenous processes such as lipid peroxidation and peptide- and amino acid nitrosation to reactive alkylating agents (malondialdehyde and alkyldiazonium ions, respectively). Recent results suggest that the characteristic DNA adducts formed by these pathways can be detected with sufficient sensitivity in blood and urine samples to enable them to be used in studies on human subjects.

Acknowledgements

This work was supported by MAFF Contract No. 1A025 and the Medical Research Council. We gratefully acknowledge the help of Dr. Gavain Sweetman for help with mass spectral analyses and Ms. Rebekah Jukes for assistance with NMR spectra.

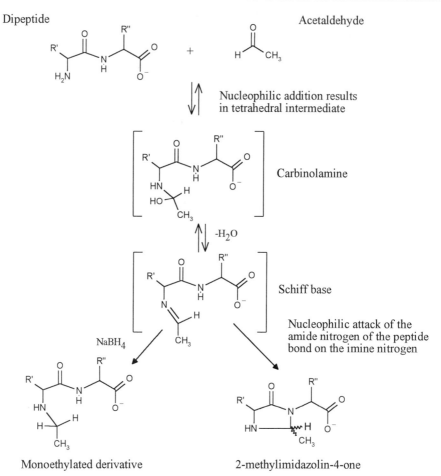

Scheme 3. *Reactions of acetaldehyde with the terminal amino group of peptides (adapted from San George and Hoberman[34])*

References

1.   IARC Scientific Publication No. 100 (1990), IARC, Lyon, France.
2.   IARC Scientific Publication No. 120 (1992), IARC, Lyon, France.
3.   D. M. Parkin, in IARC Publication No. 123 (1993), IARC, Lyon, France, Chap.1.
4.   L. N. Kolonel, A. M. Y. Nomura, T. Hirohata, J. H. Hankin and M. W. Hinds, *Am. J. Clin. Nutr.*, 1981, **34,** 2478.
5.   Consensus report in "Mechanisms of Carcinogenesis in Risk Identification" (H. Vainio, P. Magee, D. McGregor and A. J. McMichael, eds), IARC Scientific Publication No. 116, IARC, Lyon, France (1992) pp.9-54.
6.   B. N. Ames, L. S. Gold and W. C. Willett, *Proc. Natl. Acad. Sci. USA* 1995, **92,** 5258.

7.   K. A. Steinmetz and J. D. Potter, *Cancer Causes and Control,* 1991, **2**, 325.
8.   R. Saracci, *J. Intern. Med.,* 1993, **233**, 41.
9.   D. E. G. Shuker and P. B. Farmer, *Chem. Res. Toxicol.,* 1992, **5**, 450.
10. P. B. Farmer, *Clin. Chem.,* 1994, **40**, 1438.
11. P. B. Farmer and G. M. A. Sweetman, *J. Mass. Spectrom.,* 1995, in press.
12. D. E. G. Shuker and H. Bartsch, *Mutat. Res.,* 1994, **313**, 263.
13. B. C. Challis, *Cancer Surveys,* 1989, **8**, 363.
14. D. E. G. Shuker, S. R. Tannenbaum and J. S. Wishnok, *J. Org. Chem.,* 1981, **46**, 2092.
15. S. Puju, D. E. G. Shuker, W. W. Bishop, K. R. Falchuk, S. R. Tannenbaum and W. G. Thilly, *Cancer Res.,* 1982, **42**, 2601.
16. W. F. Busby, Jr., D. E. G. Shuker, G. Charnley, P. M. Newberne, S. R. Tannenbaum and G. N. Wogan, *Cancer Res.,* 1985, **45**, 1367.
17. D. E. G. Shuker, J. R. Howell and B. W. Street, in "Relevance of N-Nitroso Compounds to Human Cancer: Exposures and Mechanisms" (H. Bartsch, I. K. O'Neill and R. Schulte-Hermann, eds), IARC Scientific Publication No. 84, IARC, Lyon, France, pp. 187-190.
18. D. Anderson, B. J. Phillips, B. C. Challis, A. R. Hopkins, J. R. Milligan and R. C. Massey, *Food Chem. Toxicol.,* 1986, **24**, 289.
19. D. Anderson and S. D. Blowers, *Lancet,* 1994, **344**, 343.
20. O. Bulay, S. S. Mirvish, H. Garcia, A. F. Pelfrene and B. Gold, *J. Natl. Cancer Inst.,* 1979, **62**, 1523.
21. A. Maekaiva, T. Ogiu, C. Matsouka, H. Onedura, K. Furuta, H. Tanagawa and S. Odashima, *J. Cancer Res. Clin. Oncol.,* 1983, **106**, 12.
22. J. Zurlo, T. J. Curphey, R. Hiley and D. S. Longnecker, *Cancer Res.,* 1982, **42**, 1286.
23. H. Seto, T. Takesue and T. Ikemura, *Bull. Soc. Chem. Jpn.,* 1985, **58**, 3431.
24. L. J. Marnett, A. K. Basu, S. M. O'Hara, P. E. Weller, A. F. M. Maqsudur Rahman and J. P. Oliver, *J. Amer. Chem. Soc.,* 1985, **108**, 1348.
25. A. K. Chaudhary, M. Nokubo, G. R. Reddy, S. N. Yeola, J. D. Morrow, I. A. Blair and L. J. Marnett, *Science (Washington),* 1994, **265**, 1580.
26. M. Hadley and H. H. Draper, *Lipids,* 1990, **25**, 82.
27. S. Agarwal and H. H. Draper, *Free Rad. Biol. Med.,* 1992, **13**, 695.
28. H. K. Jajoo, P. C. Burcham, Y. Goda, I. A. Blair and L. J. Marnett, *Chem. Res. Toxicol.,* 1992, **5**, 870.
29. C. E. Vaca, P. Vodicka and K. Hemminki, *Carcinogenesis,* 1992, **13**, 593.
30. M.-Y. Yang and J. G. Liehr, *Arch. Biochem. Biophys.,* 1995, **316**, 38.
31. M. D. Gross, S. M. Gapstur, J. D. Belcher, G. Scanlan and J. D. Potter, *Alcohol. Clin. Exp. Res.,* 1992, **16**, 1093.
32. V. J. Stevens, W. J. Fantl, C. B. Newman, R. V. Sims, A. Cerami and C. M. Peterson, *J. Clin. Invest.,* 1981, **67**, 361.
33. L. Itälä, K. Seppä, U. Turpeinen, and P. Sillanaukee, *Anal. Biochem.,* 1995, **224**, 323.
34. R. San George, and H. D. Hoberman, *J. Biol. Chem.,* 1986. **261**, 6811.
35. K. P. Braun, R. B. Cody, D. R. Jones, and C. M. Peterson, *J. Biol. Chem.,* 1995. **270**, 11263.
36. R. C. Lin, J. B. Smith, D. B. Radtke, and L. Lumeng, *Alcohol. Clin. Exp. Res,* 1995, **19**, 314.

The Use of a Flow Cell Bioreactor to Monitor Chronic Exposure

Jo McBride, Martin Dyer, John Pugh, Sarah Oehlschlager, and Bryan Hanley

CSL FOOD SCIENCE LABORATORY, NORWICH RESEARCH PARK, COLNEY, NORWICH NR4 7UQ, UK

1 INTRODUCTION

A range of biological systems is available to assess the consequences of exposure to bioactive compounds. With all model systems, there are significant limitations to the applicability of the approach, many of which are recognised and at least some of them are starting to be resolved. The range of possible test systems depends, in part, on the type of biological activity being assessed. However, for **dietary** constituents with an unknown biological profile, the possibilities are a little more restricted. The principal requirements for a test system are firstly that it should be possible using such a system to mimic human chronic exposure to the complex mixture of bioactive constituents in food over a long period of time; and secondly that the indices chosen to enable an assessment of the likely biological outcome from exposure should be relevant to the human situation either by measurement of similar indices in the population or through an understanding of the fundamental biological mechanisms associated with the chosen indices of effect. *In vitro* cell culture systems are available which can be used to respond to these challenges.

The system selected for the *in vitro* study of chronic exposure to dietary toxins was human cells grown in a flow cell bioreactor. The advantages specific to this approach are: the use of cells derived from a human tissue source; the growth of cells in a *continuous* fashion which is closer to the *in vivo* situation than using that of standard flask culture in which growth is *discontinuous*, and a significantly reduced cost in comparison to animal studies. Additionally, the use of a stable cell line improves the reproducibility of the results and minimises variation.

There are, however, obstacles which must be overcome if this approach is to be considered a valid biomimetic system. For example, culture and growth conditions must be developed such that a range of human cells can be grown in the bioreactor. Conventional approaches using flask culture systems have led to the development of culture conditions specific to that environment, and these conditions may not be suitable or optimal for cells in the bioreactor environment.

The major targets of this work were therefore, firstly to develop the flow cell bioreactor such that it can be used to culture human cells for extended periods of time, and secondly to measure parameters of cell growth which can be used to determine the biological state of the cells in the bioreactor.

2 THE BIOREACTOR

The bioreactor used in these studies, (Tecnomouse, patent pending Endotronics Inc and Tecnomara AG) is supplied by Integra Biosciences UK. The original application for this apparatus is the production of antibodies from hybridomas. We have investigated the possibility of using the system for continuous culture of attaching human cell lines. The bioreactor can contain up to five hollow fibre cassettes through which medium is constantly circulated. Cells grow inside the cassette on the outside of the fibres in the extra capillary (EC) space. Nutrients are delivered via the intracapillary (IC) space and enter the EC space by dialysis. The cassettes are maintained at a temperature of 37°C and in a modified atmosphere (5% CO_2).

Human secondary cells were grown in a suitable medium in 75cm^2 tissue culture flasks, harvested and cryopreserved until sufficient cells were available for a single inoculation into the bioreactor. The minimum number of cells required for inoculation is in the range 3 x10^7 to 3 x10^8. Cells seed around the hollow fibres, and also form free spherical micro-colonies. If an attaching cell line is used the population concentrates around the inoculation port. This can be partly overcome by inoculating through both EC ports and briefly shaking the cassette to ensure a more even distribution. For routine examination samples can be removed from the EC space using a syringe. In our laboratory a recently harvested cassette with a starting density of 1.09 x10^8 cells achieved a final density of 5.42 x10^8 cells over a period of 7 weeks.

3 CELL GROWTH PARAMETERS

Chang liver cells were inoculated into the bioreactor cassette at a seeding density of 3 x 10^7 in 5mls of MEM medium (+10%) Bovine serum. Medium (MEM, 2.0l) was constantly circulated through the hollow fibres (IC space) at rates between 100-150ml hr^{-1} (this medium was replaced weekly). Flow rate was determined, in part, by cell density, and could be increased with increasing cell number to ensure that sufficient nutrients reached the whole population. In addition, the medium in the growth chamber of the cassette (the outside of the fibres), which includes serum, was replaced every 3 days. A solution of antibiotics and an antimycotic (penicillin, streptomycin, amphotericin, Sigma A9909) was routinely added to the circulating medium in order to minimise the risk of bacterial/fungal contamination.

Three growth parameters were measured:-

- Cell viability by trypan blue staining
- Glucose uptake from the medium as a measure of metabolic activity
- Lactate dehydrogenase (LDH) as a measure of cell death.

3.1 Cell Viability

Cells were removed from the bioreactor either by a brief trypsinisation followed by syringe withdawal (Chang A), or by syringe withdrawal alone (Chang B). The cells were stained with trypan blue, and counted in a haemocytometer.

3.2 Lactate Dehydrogenase Activity

As cells become non-viable they release a number of cellular components into the growth medium including the enzyme lactate dehydrogenase (LDH). The levels of this enzyme present in medium from the EC space give an indirect measurement of the extent of cell death since the previous sampling and confirmation of continued cell division. LDH activity was determined using an assay kit (Sigma LD L Procedure No 228-UV, The quantitative kinetic determination of LDH in serum at 340nm) which measures the production of NADH spectrophotometrically in a reaction catalysed by LDH.

3.3 Glucose Uptake and Lactate Production

The circulating medium contains glucose, and the gradual removal of this energy source provides a measure of the metabolic state of the cells in the bioreactor. The level of residual glucose in the medium was measured enzymatically in a coupled assay using hexokinase and glucose-6-phosphate dehydrogenase which react sequentially with glucose to produce, as a side product from an essential co-factor, NADPH. The latter can be measured spectrophotometrically. This assay was carried out using Sigma kit Glucose-HK ((procedure No 16-UV, Quantitative enzymatic (hexokinase) determination of glucose in serum or plasma at 340nm)).

The production of lactate is an indicator of glucose metabolism in the cells since it represents an accumulated metabolic end point. Lactate can be assayed in a coupled system whereby lactate is converted to pyruvate and hydrogen peroxide by lactate oxidase. The hydrogen peroxidase thus formed reacts with a chromogenic substrate in the presence of horseradish peroxidase, and the colour change detected spectrophotometrically. The kit used was Sigma Lactate (Quantitative enzymatic determination of lactate in plasma at 540nm, procedure No 735).

4 Results and Discussion

The Chang liver cells have been in continuous culture in the bioreactor since August 1994 (Chang Liver A), and the results of the cumulative data for an eight month period are shown in Table 1. The values for LDH indicate that there is a constant cycling of cells in the cassette with a release of LDH being an indication of death, while the residual glucose value establishes that glucose is taken up at a relatively consistent rate by the live population. The value for lactate production is more variable, and this may mean that some anaerobic stress is being placed on the cassette as the cell number increases, and that the restriction placed on population expansion is governed by gas exchange, and the transport of nutients across the membrane. It is not possible to assess the total number of living cells in the cassette except by sacrifice and the assessments made by following rapid trypsinisation did not appear to show a very high cell viability (ca 10%).

Table 1 *Cumulative data for metabolic markers, LDH, Lactate and Glucose uptake for two populations (A and B) of Chang Liver cells cultured in the bioreactor*

Chang A	MEAN	ERROR %	N
LDH(U/L) Log 10	3.24	14	17
Lactate (mMol)	2.57	27	17
Residual Glucose (mMol) of 25 mMol)	19.59	10	17
Cell Viability (%)	11.94	70	13
Cell death (%)	88.06	9.5	13
Chang B			
LDH(U/L) Log 10	3.31	10.5	18
Lactate (mMol)	1.76	31	17
Residual Glucose (mMol) of 25 mMol)	21.30	10	17
Cell Viability (%)	10.55	60	5
Cell death (%)	89.45	7.1	5

A second study on Chang liver cells (Chang Liver B) was initiated to further define the growing conditions. The initial investigation (Chang Liver A) involved significant disruption of the culture by trypsinisation to remove cells. In this case cells were withdrawn using negative pressure through a syringe, and were not subjected to trypsin activity. Cells withdrawn in this manner can be observed as micro-colonies, (small spheroids of cells, which appear to contain a central core of dead material which stains positively with trypan blue). When seeded into a tissue culture flask, micro-colonies disaggregate over a period of seven days and resume typical monolayer growth.

The cumulative data for a five month period are shown in Table 1. The values for LDH, lactate and residual glucose and cell viability are not significantly different between cassettes A and B, but the standard errors are less for B in each case. However the mean value for lactate decreased in cassette B but the corresponding error showed an increase of 5%. The values recorded for the second cassette (Chang B) may indicate that the disruption of cassette A through trypsinisation did not result in a better recovery of viable cells, but did have the effect of perturbing the cell population. If this is the case then the values recorded for lactate in cassette B probably reflect the rate limiting effects of gas exchange across the cassette membrane more accurately. Using these criteria it can be seen from Figure 1 that the first peak of cell numbers occurred between months 3 and 4 for both cassettes, and that a gradual recovery is evident for cassette A from that time on.

The low number of viable cells recovered from the cassettes whether by trypsinisation or direct withdrawal reflects the fact that Chang Liver cells are a very adherent cell line and the living cells are difficult to dislodge from the hollow fibre substrate. The sample

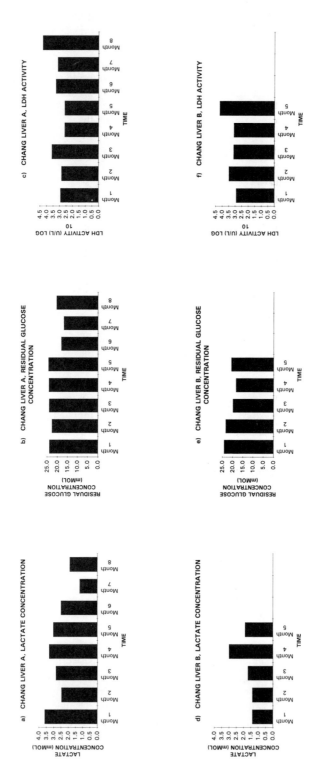

Figure 1 a) b) and c) *Chang Liver A, Culture cassette seeded with Chang Liver Cells, growth profile over an eight month period measured by lactate production, glucose uptake and LDH activity d) e) and f) Chang Liver B, a similar cassette seeded with Chang Liver Cells measured using the same growth parameters*

then becomes enriched for dead cells which dislodge more easily and were found in consistently large numbers.

5 Conclusion

We have demonstrated that it is possible to maintain a population of secondary human liver cells in a flow cell bioreactor for periods of up to 6 months. The cassette contains ~10^8 cells in a 10ml volume which is constantly supplied with a re-circulating medium that passes through the hollow fibre membrane supplying nutrients and carrying away metabolites. The results from the assay of LDH and lactate indicate that a dynamic population is maintained in the culture cassette for a period of several months, and thus would be amenable as a long term exposure system.

The bioreactor can be used in two ways, firstly as a chronic exposure system to monitor chronic versus acute effects of dietary toxins. Changes in basal metabolism, and biomarkers can be measured and selected as anchor points for toxin induced change. The relevance of the biomarkers chosen will need to reflect an understanding of the biological mechanisms operating in the bioreactor. Secondly, while this cannot have a direct relationship with the whole human target, the experience gained from the operation and manipulation of such a system should permit the articulate design of human studies, and provide some insight into the measurements that can be made in the complete living human. The validity of the cell system can be examined by attempting to measure similar markers in clinical samples, and through dietary and epidemiological studies.

Acknowledgement: This work was supported by the Ministry for Agriculture, Fisheries and Food

Application of Mechanistic Studies of Fecapentaene-12 Genotoxicity in the Development of a Specific Biomarker of Exposure to this Compound

S. M. Plummer,[1] M. Hall,[1] and S. P. Faux[2]

[1] CENTRE FOR MECHANISMS OF HUMAN TOXICITY, HODGKIN BUILDING, PO BOX 138, LANCASTER ROAD, LEICESTER LE1 9HN, UK

[2] INSTITUTE OF OCCUPATIONAL HEALTH, UNIVERSITY OF BIRMINGHAM, EDGBASTON, BIRMINGHAM B15 2TT, UK

1 INTRODUCTION.

Although the origins of colon cancer are unknown, there is epidemiological evidence that the diet is a major risk factor in this disease (1). High fat and meat and low fibre intake have been associated with increased risk. The mechanism , however, whereby these dietary components alter risk is unclear. One hypothesis is that dietary factors might alter the production and bioavailability of carcinogenic compounds in the stool (2). This hypothesis stems from the fact that colon cancer occurs mostly in the sigmoid colon where faecal contents are concentrated. As most carcinogenic agents are mutagens, Bruce and coworkers decided to investigate the mutagenic potential of human faecal extracts. These investigators found that organic solvent extracts of human faeces from individuals consuming a 'Western' diet were highly mutagenic, and most of this mutagencity resided in a single fraction (3). Chemical analysis established that this fraction, which accounted for 90% of mutagenic activity in faeces (4), contained enol-ether lipids which were called fecapentaenes (5). Subsequent work established that fecapentaenes were synthesised in the colon from precursors of unknown origin called plasmalopentaenes by phospholipase and lipase cleavage in the membranes of certain strains of *Bacteroides* (6). This process required the presence of bile acids which act as solubilising agents.

Fecapentaenes are mutagenic at nanomolar concentrations in the Ames test (7) , and are comparable in potency with compounds like aflatoxin B_1 and benzo(a)pyrene except that, unlike these compounds, fecapentaenes do not require metabolic 'activation'. Fecapentaene-12 (fec-12), a commercially available model fecapentaene and the most abundant form in faecal extracts, is active in a range of short term assays for genotoxicity in normal human fibroblasts including DNA single strand breaks (SSB), unscheduled DNA synthesis (UDS), sister chromatid exchange, and mutations in the HGPRT locus (8). In vivo studies in rodents have shown that fec-12 will induce DNA SSB and UDS in the colon when administered intrarectally (9). Fec-12 was not carcinogenic in the rat colon but weakly active at other sites when given intrarectally and induced tumours in neonatal mice when given i.p. (10, 11, 12). Fecapentaenes display tumour promoting activity in a rat colon carcinogenesis model system with methylnitrosourea as the initiator (13).

Fecapentaenes are highly reactive towards oxygen and break down rapidly in oxygenated aqueous solutions. The structural features which confer this reactivity, namely the fully conjugated 12 carbon chain, together with the enol ether oxygen, are also important requirements for their activity as mutagens in the Ames Test (14). This structure activity data, coupled with the observations that mutagencity to TA100 bacteria which detect base substitution mutagens requires the presence of oxygen (15), and that strains of bacteria which were designed to detect oxidative mutagens are more sensitive to the effects of these compounds (7), led to the hypothesis that they act as mutagens through oxidative mechanisms (16). This is supported by the observations that fec-12 induces 8-

hydroxydeoxyguanosine (8-OHdG) in isolated DNA (17) and forms oxygen and alkyl radicals in oxygenated aqueous solutions (18). However this is not the whole story since fecapentaenes are also mutagenic to TA 98, which detect frameshift mutagens in the absence of oxygen (15) . Chemical studies have shown that fecapentaenes have alkylating properties as a result of their ability to protonate and form carbocations (19). Fecapentaenes, therefore, may have two different mechanisms of action: (1) oxidative decomposition to form reactive oxygen species and aldehydes; and (2) direct alkylation of DNA. Evidence in support of fecapentaene-related DNA adduct formation has come from DNA postlabelling studies (18).

Epidemiological studies indicate that higher levels of faecal mutagencity are positively associated with risk for colon cancer. However mutagencity does not always correlate with levels of fecapentaenes in faecal extracts due to the presence of inhibitory agents (20). Case control studies have shown that fecapentaene excretion is negatively associated with risk for colon adenomas and carcinomas (21, 22), and recent dietary studies have shown that vegetarians, who are at decreased risk for colon cancer compared to omnivores, excrete more fecapentaenes than omnivores (23). This suggests that increased excretion of fecapentaenes may be a protective factor. These studies illustrate the difficulties in interpretation of fecapentaene excretion data in the absence of an indicator of fecapentaene bioavailability in the target tissue, i.e. the colon epithelium. There are many agents in the colon lumen, e.g. bile acids and calcium, which increase and decrease fecapentaene absorption in model systems (24). Since it is difficult to control for these factors, it is impossible to estimate internal target exposure of fecapentaenes by measuring lumenal concentrations. A biomarker of fecapentaene exposure is therefore required. If it is hoped to eventually use this biomarker for cancer risk assessment purposes, the marker should be relevant to the genotoxic mechanism of action of fecapentaenes. Towards that end, this paper describes our recent investigations of the genotoxic mechanism of action of fec-12 in human cells.

2. MECHANISM (S) OF ACTION OF FECAPENTAENE-12

We wanted to investigate further the oxidative mechanism of fec-12 genotoxicity in human cells for two reasons. Firstly, several studies have shown that a significant proportion of inactivating point mutations in the tumour suppressor genes P53 and APC in human colon cancer are cytosine to thymidine transitions occurring mainly at CpG dinucleotides (25, 26, 27). This particular type of mutation can occur spontaneously in cells through hydrolytic deamination of cytosine, but occurs much more rapidly in the presence of reactive oxygen species (28). Secondly, de Kok and coworkers observed that isolated preparations of peroxidative enzymes, including prostaglandin H synthase (PGHS), potentiate the generation of reactive oxygen species by fec-12 (29). These findings suggest that oxidative mechanisms of DNA damage, a component of fecapentaene damage to isolated DNA, may also be important in the induction of critical mutations in colon carcinogenesis, and that cellular metabolism of fec-12 may potentiate this damage. Therefore the question arose whether oxidative DNA damage was induced in human cells exposed <u>in vitro</u> to fecapentaenes, and if this damage was potentiated by PGHS. The possibility that PGHS might be involved in this potentiation was particularly interesting in view of recent epidemiological data showing an inverse dose response relationship between the intake of aspirin , a PGHS inhibitor, and risk for colon cancer.

2.1 Formation of 8-OHdG in isolated DNA or HeLa cells exposed to fec-12.

The formation of oxidative DNA base damage was investigated in HeLa cells using HPLC with electrochemical detection (30). Fec-12 exposures at sub-toxic doses were shown to induce a dose dependent increase in the formation of 8-OHdG in the cells following a 1hr exposure at 37°C (31) (Table 1.). It was also found that exposure of isolated DNA to fec-12 produced increases in 8-OHdG levels which were similar to those seen in cells. The concentrations of fec-12 required to produce these increases, however, were an order of magnitude higher than those used in the cellular experiments, suggesting that potentiation of the ability of fec-12 to induce oxidative damage was occurring in cells. To ascertain whether the cellular potentiation of fec-12 DNA damage was mediated by PGHS, we investigated whether non-steroidal anti-inflammatory agents, which are known to inhibit PGHS, might also inhibit the induction of oxidative DNA damage by fec-12 in HeLa cells. It was found that exposure of HeLa cells to fec-12 in combination with the PGHS inhibitor indomethacin blocked the increase in 8-OHdG levels in DNA seen after exposure to fec-12 alone. These results indicate firstly that fec-12 does induce oxidative DNA damage in human cells, which corroborates earlier work in bacteria, and adds further support to the theory that they act in part through the production of reactive oxygen species, and that induction of oxidative DNA damage in human cells is potentiated by PGHS. The latter finding suggested that this enzyme is involved in 'activating' fec-12 to generate reactive oxygen species in cells.

Table 1. *Formation of 8-OHdG in HeLa cell DNA following exposure to fec-12 .*

Treatment	Moles 8OHdG/ . Mol dG (%)	% Viability
Control	0.0047±0.0012	93.0
Fec-12 10μM	0.0050±0.0026	90.0
Fec-12 50μM	0.0111±0.0013	87.0
Fec-12 100μM	0.0137±0.0006	85.0
Fec-12 (50μM)+Indo (100μM)	0.0043±0.0023	90.2

2. 2 Which component of PGHS activity is responsible for the potentiation of fec-12 genotoxicity?

Prostaglandin H synthase is a bifunctional enzyme which incorporates separate cyclooxygenase and peroxidase activities in the same protein. The cyclooxygenase inserts oxygen in the unsaturated hydrocarbon chain of arachidonate to generate a hydroperoxide, PGG2. The peroxidase reduces these hydroperoxides to the corresponding alcohol, PGH2 (32). To investigate the role of these two activities, oxygen uptake was examined as an indicator of interaction between a substrate and PGHS, using fec-12 with purified preparations of PGHS in the presence or absence of selective inhibitors of the cylooxygenase or peroxidase activities of the enzyme. Furthermore, the effects of the same inhibitors on the induction of DNA single strand breaks by fec-12 in HeLa cells were measured to assess the role of the two PGHS components in fec-12 genotoxicity.

To inhibit the cyclooxygenase, indomethacin was selected as it competes with the substrate, arachidonate, for the active site (33). To inhibit the peroxidase , methylphenyl sulphide (MPSI) was chosen, which acts as a reducing cosubstrate thus preventing peroxidase cooxidation reactions (34). Unlike other reducing substrates which undergo two sequential one electron transfers with the concomitant production of radicals, MPSI donates two electrons simultaneously by reacting with oxygen in the ferroxyl active site of the peroxidase, and is converted to methylphenyl sulphoxide, a radical scavenger . Therefore MPSI blocks peroxidase mediated cooxidation of fec-12 by preventing the formation of free radicals which propagate these reactions.

2.3 Oxygen uptake

Oxygen uptake occurs with fec-12 in the absence of PGHS, which reflects the reactivity of this compound with oxygen. However, oxygen uptake with fec-12 was markedly potentiated by the presence of PGHS. Indomethacin has an inhibitory effect on the initial rate of oxygen uptake while MPSI causes a more marked inhibition of oxygen uptake than indomethacin. The two agents in combination have an additive effect. Oxygen uptake with fec-12 PGHS is several fold greater than that seen with the endogenous substrate arachidonate (Figure 1.).

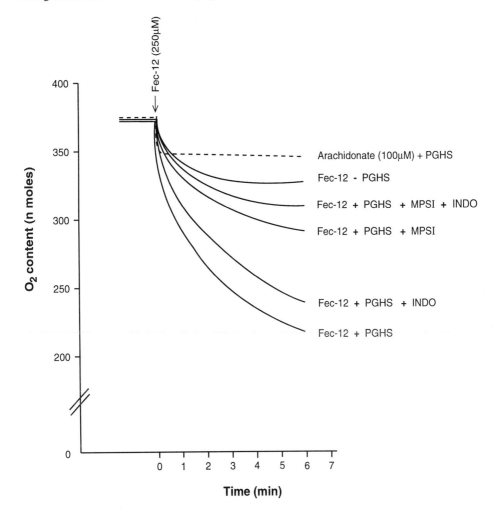

Figure 1. *Oxygen uptake by fec-12 with or without PGHS in the absence or presence of inhibitors.*

2.4 PGHS mediated induction of DNA SSB in HeLa cells exposed to fec-12 in the presence or absence of indomethacin or MPSI.

We investigated the role of the two PGHS activities in the genotoxic effects of fec-12 in cells (HeLa) by measuring the effects of the inhibitors on the formation of fec-12 induced DNA SSB . DNA SSB were measured by two methods: Fluorescence Alkaline DNA Unwinding (FADU), which measures the amount of double stranded DNA remaining in a sample by ethidium fluorescence (35), and the 'comet' assay which measures DNA SSB in individual cells by quantitating the migration of DNA after cells are embedded in an agarose gel and subjected to alkaline electrophoresis (36).

Fec-12 exposure caused increases in the amount of SSB in cellular DNA but these increases were blocked by the presence of indomethacin and aspirin , suggesting that the cyclooxygenase has a role in fec-12 genotoxicity (Figure 2.).

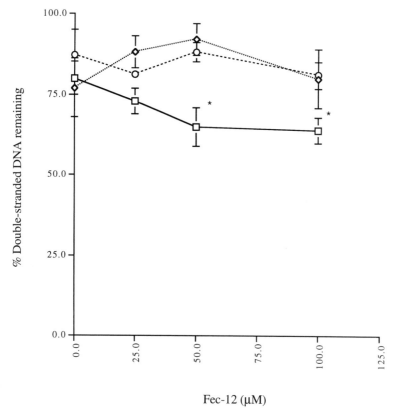

Fec-12 (μM)

Figure 2. *Induction of DNA SSB in HeLa cells by fec-12 in the absence or presence of indomethacin or aspirin, as measured by FADU.*

Fec-12 exposure also caused a dose dependent increase in DNA SSB, which is reflected by an increase in comet tail length. MPSI blocked the increase in DNA SSB induced by fec-12 to a similar extent with all three concentrations tested (Figure 3.).

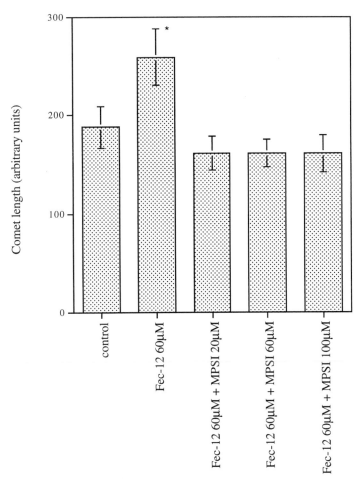

Figure 3. *Effect of methylphenylsulphide (MPSI) on the increase in comet 'head' to 'tail' length induced by exposure to fec-12 .*

These data indicate that both cyclooxygenase and peroxidase activities of PGHS mediate the genotoxicity of fec-12 in HeLa cells. A possible explanantion for these results is that indomethacin , which selectively inhibits PGHS cyclooxygenase by blocking access of the substrate (arachidonate) to the active site, could also block the cyclooxygenase from forming highly reactive fec-12 hydroperoxides which would be genotoxic (Figure 4.). The oxygen uptake data with fec-12 and purified preparations of PGHS showing inhibition by indomethacin provide support for this theory.

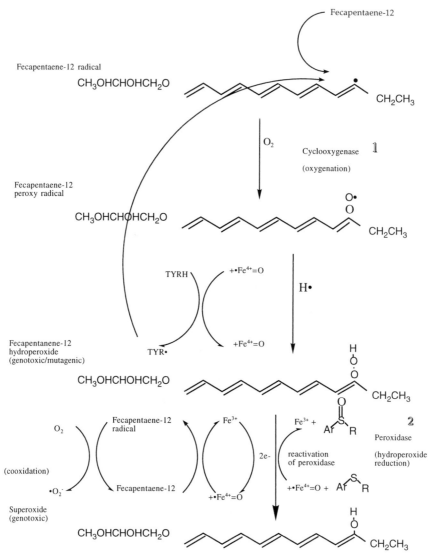

Figure 4. *Hypothetical mechanisms of fec-12 'activation' by prostaglandin H synthase. ArSOR (methylphenylsulphoxide), ArSR (methylphenylsulphide), TYRH (tyrosine active site of cyclooxygenase), TYR• (tyrosyl radical).*

There are two ways in which MPSI could modulate fec-12 genotoxicity. It could either block peroxidase cooxidation of fec-12, which generates superoxide, or it could potentiate the activity of the peroxidase by maintaining the enzyme in an active oxidation state, thus maximising its ability to reduce putative fec-12 hydroperoxides to less reactive alcohols. Both these actions could contribute to the inhibitory effects of this compound on fec-12 genotoxicity. However, the latter reaction would be more important for the biological activity of fec-12 in cells because superoxide, generated through peroxidase cooxidation, is not a potent genotoxicant.

3. MUTAGENESIS

To further investigate the biological significance of fec-12 activation by PGHS, we examined the mutagenic potential of fec-12 in the presence of microsomal PGHS preparations . This was done via mutagenicity studies in the Lac I target gene of Big Blue™ rat fibroblasts (BBRF) . As these cells contain a transgene, LacI, which is incorporated into chromosomal DNA via a shuttle vector , the gene can be rapidly excised, packaged into a phage and analysed for mutations at the molecular level (37). This system therefore facilitates the rapid gathering of mutational information in mammalian cells.

In these experiments, BBRF cells were exposed to fec-12 alone or with HeLa cell or human colon microsomes in the presence or absence of indomethacin. Since the source of microsomal activation was outside the target cells, it is unlikely that much contribution to the effects of fec-12 could come from PGHS peroxidase cooxidation mediated production of superoxide since superoxide, which is charged, cannot cross cell membranes.

3. 1 Mutagencity of fec-12 in Big Blue™ rat fibroblasts.

BBRF cells were relatively resistant to the mutagenic effects of fecapentaenes alone when compared to human fibroblasts (8). However, microsomes from HeLa cells potentiated the mutagencity of fec-12 in these cells 4 fold (Figure 5.).

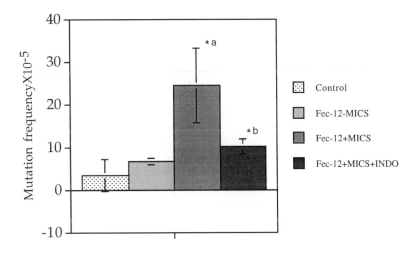

Figure 5. *BBRF cell mutagenicity of fec-12 with or without HeLa cell microsomes in the absence or presence of indomethacin.*

Indomethacin markedly inhibited the mutagenic effect of fec-12 in the incubations with microsomes but not with fec-12 alone. This suggests that the cyclooxygenase component of microsomal PGHS generates mutagenic intermediates from fec-12 and that these intermediates can gain access to the nucleus from outside the cell. To assess the biological significance of this finding for the target organ, the experiments were repeated with human colon epithelial microsomes; similar results were obtained (38). These findings are biologically relevant because the concentrations of fec-12 used in these experiments are similar to those found in the human colon (39), and PGHS has been shown to be active in human colon microsomes (40).

3.2 Mutation spectrum of fec-12 in Big Blue™ rat fibroblasts

To gain information about the mechanism whereby these mutations were being induced, we characterised, by DNA sequencing, a series of the fec-12 plus HeLa cell microsomes mutations using PCR and automated sequencing. We found that almost all of the mutations occurred at G.C base pairs and that most (75%) were base pair substitutions. Of these point mutations, there was a predominance (70%) of G.C to T.A transversions. These mutations can be induced by at least two different types of DNA damage. One is the formation of bulky guanine adducts such as benzo(a)pyrene N7 adducts, which result in polymerase error according to the 'A' rule (41). The other is the formation of 8-OHdG, which causes mispairing with adenine (42). Since we have shown that fec-12 will induce 8-OHdG in human cells and that this, together with the induction of mutations in BBRF cells, is inhibited by indomethacin, it seems likely that these mutations could be induced by the formation of 8-OHdG. It is not possible to positively identify the oxygen radical(s) which mediate the formation of 8-OHdG since both OH radicals and singlet oxygen will do this. However since the mutations were confined almost exclusively to G.C base pairs, singlet oxygen appears a more likely candidate since it degrades only guanine (43) while OH radicals attack all DNA bases (28).

4. HOW THIS DATA MIGHT BE USED IN THE CHOICE OF AN APPROPRIATE BIOMARKER OF FECAPENTAENE EXPOSURE

These studies indicate that fec-12 is activated by enzymes in the target tissue (colon) to intermediates which cause DNA damage and mutations via mechanisms which are important in the process of human colon carcinogenesis. The data would indicate that it is the hydroperoxides of fec-12 formed by the cyclooxygenase component of PGHS which precipitates the relevant genotoxicity. Hydroperoxides can cause DNA damage via fenton-like reactions forming alcoxyl radicals (44). These radicals react with oxygen to form singlet oxygen and aldehydes (45). Singlet oxygen attacks guanine bases in DNA via addition reactions to form guanine hydroperoxides, which in turn undergo reduction to form 8-OHdG (46). Since 8-OHdG can also be formed in DNA through the attack of hydroxy radicals in the 8 position of guanine, the measurement of 8-OHdG does not provide a specific biomarker of fec-12 exposure. DNA adducts formed by reactions with fecapentaene derived aldehyde moieties may, however, provide a specific indicator of fec-12 exposure because they would retain structural characteristics unique to fecapentaene.

Povey et al. have shown by DNA 32-P postlabelling analysis that DNA reacted in isolation with fec-12 contains aldehyde-like DNA adducts (18). Since DNA 32 P postlabelling is very sensitive and can be applied to as little as one microgram of DNA, it is possible that this method could be applied in the detection of fec-12 adducts in human colon biopsy specimens which are taken during routine screening for colon cancer.

5. HOW MIGHT THE INFORMATION GAINED FROM THESE STUDIES BE USED TO PREVENT COLON CANCER?

Since PGHS appears to be important in the genotoxicity of fec-12 in target tissues, it is possible that inhibitors of this enzyme might be useful for colon cancer prevention by blocking fec-12 genotoxicity. In this regard, it is interesting that several recent epidemiological studies have shown negative correlation between colon cancer risk and intake of the non-steroidal anti-inflammatory agent (NSAID) aspirin (47). Sulindac, another NSAID , has been shown to reverse the onset of polyps in FAP pateints (48). Many animal studies have demonstrated that NSAIDs will inhibit induction of colon cancer induced by various model carcinogens, e.g. N-methylnitrosourea (49). Gaining insight into the mechanism of PGHS activation of fec-12 could lead to the development of new approaches to colon cancer prevention. Recently another isoform of PGHS was discovered called PGHS2 (50). This enzyme is regulated differently from PGHS1 and is induced by various mitogenic stimuli eg oxidative stress. PGHS2 appears to be expressed selectively in human colon tumours compared to normal colon epithelium (51). This suggests that expression of PGHS2 might be relevant to human colon carcinogenesis. It will be interesting to see whether both form of PGHS can activate fec-12. Design of agents which specifically target PGHS2 may provide useful chemopreventive agents without the unwanted side effects of non-selective inhibitors like aspirin.

The diet also contains naturally occurring PGHS inhibitors e.g. curcumin. Measurement of fec-12-DNA adducts as a biomarker of fec-12 exposure in dietary intervention studies could lead to a greater understanding of how the diet modulates fecapentaene genotoxicity and may enable the development of novel approaches to dietary chemoprevention of colon cancer.

References

1. B. Armstrong and R. Doll, *Int. J. Cancer* **15**, 617-631 (1975).
2. P. Correa and W. Haenszel, *Advances in Cancer Research* **26**, 1-141 (1978).
3. W. R. Bruce, A. J. Varghese, R. Furrer and P. C. Land, *A Mutagen in the Feces of Normal Humans*. H. H. Hiatt Watson, J.D., and Winsten, J.A., Eds., Origins of Human Cancer (Cold Spring Harbour Press, Cold Spring Harbour, NY, 1977).
4. P. Dion and R. Bruce, *Mutation Research* **119**, 151-160 (1983).
5. I. Gupta, J. Bapista, W. R. Bruce, C. T. Che, R. Furrer, J. S. Gingerich, A. A. Grey, L. Marai, P. Yates and J. J. Krepinsky, *Biochemistry* **22**, 241-245 (1983).
6. R. L. Van Tassell, T. Piccariello, D. G. I. Kingston and T. D. Wilkins, *Lipids* **24**, 454-459 (1989).
7. R. D. Curren, D. L. Putman, L. L. Yang, S. R. Haworth, T. E. Lawlor, S. M. Plummer and C. C. Harris, *Carcinogenesis* **8**, 349-352 (1987).
8. S. M. Plummer, R. G. Grafstrom, L. L. Yang, R. D. Curren, K. Linnainmaa and C. C. Harris, *Carcinogenesis* **7**, 1607-1609 (1986).
9. M. J. Hinzman, C. Novotney, A. Ullah and A. M. Shamsuddin, *Carcinogenesis* **8**, 1475-1479 (1989).
10. J. M. Ward, T. Anjo, L. K. Ohannesian, D. E. Devor, P. J. Donovoan, G. T. Smith, J. R. Henneman, A. J. Streeter, N. Konishi, S. Rehm, E. J. Reist, W. W. I. Bradford and J. M. Rice, *Cancer Lett.* **42**, 49-59 (1988).
11. J. H. Weisberger, R. C. Jones, C.-X. Wang, J.-Y. C. Backlund, G. M. Williams, D. G. I. Kingston, R. L. Van Tassell, R. F. Keyes, T. D. Wilkins, P. P. de Witt, M. Van der Steeg and A. Van der Gen, *Cancer Lett.* **49**, 89-98 (1990).
12. A. M. Shamsuddin, A. Ullah, A. Baten and E. Hale, *Carcinogenesis* **12**, 601-607 (1991).
13. M. Zarkovic, X. Qin, Y. Natatsuru, H. Oda, T. Nakamura, A. M. Shamsuddin and T. Ishikawa, *Carcinogenesis,* **14**, 1261-1264 (1993).
14. C. E. Voogd, L. B. J. Vertegaal, M. van der Steeg, A. van der Gen and G. R. Mohn, *Mutation Research* **243**, 195-199 (1990).

15.   S. Venitt and D. Bosworth, *Mutagenesis* **3**, 169-173 (1988).
16.   L. B. J. Vertegaal, C. E. Voogd, G. R. Mohn and A. ven der Gen, *Mutation Research* **281**, 93-98 (1992).
17.   M. Shioya, K. Wakabayashi, K. Yamashita, M. Nagao and T. Sugimura, *Mutation Research* **225**, 91-94 (1989).
18.   A. C. Povey, V. L. Wilson, J. L. Zweier, P. Kappusamu, I. K. O. O'Neill and C. C. Harris, *Carcinogenesis* **13**, 395-401 (1992).
19.   I. Gupta, K. Suzuki, W. R. Bruce, J. J. Krepinsky and P. Yates, *Science* **225**, 521-522 (1984).
20.   M. H. Schiffman, *Cancer Surveys* **6**, 653-671 (1987).
21.   P. Correa, J. Paschal and P. Pizzolato, *Fecal mutagens and colorectal polyps: preliminary report of an autopsy study*. W. R. Bruce, P. Correa, and M. Lipkin, Eds., Gastrointestinal cancer (Cold Spring Harbour Laboratory Press, Cold Spring Harbour , NY, 1981), vol. Banbury report 7.
22.   M. H. Schiffman, R. L. Van Tassell, A. Robinson, L. Smith, J. Daniel, R. N. Hoover, R. Weil, J. Rosenthal, P. P. Nair, S. Schwartz, H. Pettigrew, S. Curiale, G. Batist, G. Block and T. D. Wilkins, *Cancer Research* **49**, 1322-1326 (1989).
23.   T. M. C. M. deKok, A. van Faassen, R. A. Bausch-Goldbohm, F. ten Hoor and J. C. S. Kleinjans, *Cancer Letters* **62**, 11-21 (1992).
24.   T. M. C. M. de Kok, M. L. P. S. van Iersel, F. ten Hoor and J. C. S. Kleinjans, *Mutation Research* **302**, 103-108 (1993).
25.   M. Hollstein, D. Sidransky, B. Vogelstein and C. C. Harris, *Science* **253**, 49-53 (1991).
26.   S. M. Powell, N. Zilz, Y. Beazer-Barclay, M. Bryan, S. R. Hamilton, S. N. Thibodeau, B. Vogelstein and K. W. Kinzler, *Nature* **359**, 235-237 (1992).
27.   M. Michiko, M. Konishi, R. Kikuchi-Yanoshita, M. Enomoto, T. Igari, K. Tanaka, M. Muraoka, H. Takahashi, Y. Amada, M. M. Y. Fukayama, T. Iwama, Y. Mishima, T. Mori and M. Koike, *Cancer Research* **54**, 3011-3020 (1994).
28.   F. Hutchinson, *Prog. Nucleic Acid Res. Mol. Biol.* **32**, 414-419 (1981).
29.   T. C. M. DeKok, J. M. S. Van Maanen, J. Lankelma, F. Ten Hoor and J. C. S. Kleinjans, *Carcinogenesis* **13**, 1249-1255 (1992).
30.   S. P. Faux, J. E. Fransis, A. G. Smith and J. K. Chipman, *Carcinogenesis* **13**, 247-250 (1992).
31.   S. M. Plummer and S. P. Faux, *Carcinogenesis* **15**, 449-453 (1994).
32.   W. L. Smith and L. J. Marnett, *Biochim. Biophys. Acta* **1083**, 1-17 (1991).
33.   R. J. Kulmacz and W. E. M. Lands, *J. Biol. Chem.* **260**, 12572-12578 (1985).
34.   P. Ple and L. J. Marnett, *J. Biol. Chem.* **24**, 13989-13993 (1989).
35.   H. C. Birnboim and J. J. Jevcak, *Cancer Research* **41**, 1889-1892 (1981).
36.   N. P. Singh, M. T. McCoy, R. R. Tice and E. L. Schneider, *Experimental Cell Research,* **175**, 184-191 (1988).
37.   D. L. Wyborski, S. Malkhosayan, J. Moores, M. Perucho and J. Short, *Mutation Research* **334**, 161-165 (1995).
38.   S. M. Plummer, *British Journal of Cancer* **71**, 27 (1995).
39.   M. H. Schiffman, R. L. Van Tassell, A. W. Andrews, J. Daniels, A. Robinson, L. Smith, P. P. Nair and T. D. Wilkins, *Mutation Research* **222**, 351-357 (1989).
40.   T. J. Flammang, Y. Yamazoe, R. W. Benson, D. W. Roberts, D. W. Potter, D. Z. J. Chu, N. L. Lang and F. F. Kadlubar, *Cancer Research* **49**, (1989).
41.   B. Strauss, S. Rabkin, D. Sagher and P. Moore, *Biochemie* **64**, 829-838 (1982).
42.   K. G. Higinbotham, J. M. Rice, B. A. Diwan, K. S. Kasprzak, Reed. C.D. and A. O. Perantoni, *Cancer Research* **52**, 4747-4751 (1992).
43.   J. T. Lutgerink, E. van den Akker, D. Pachen, E. J. Smeets, P. van Dijk, J.-M. Aubry, H. Joenje, M. V. M. Lafleur and J. Retel, *Singlet oxygen-induced DNA damage: Product analysis, studies of biological consequences and characterisation of mutations*. D. H. Phillips, M. Castenaro, and H. Bartsch, Eds., Postlabelling Methods for detection of DNA adducts (IARC, Lyon, 1993).

44. W. Chamilautrat, M. F. Hughes, T. E. Eling and R. P. Mason, *Archives of Biochemistry and Biophysics* **290**, 153-159 (1991).
45. R. Labeque and L. J. Marnett, *Biochemistry* **27**, 7060-7070 (1988).
46. T. P. A. Devasagayam, S. Steenken, M. S. W. Obendorf, W. A. Schulz and H. Sies, *Biochemistry* **30**, 6283-6289 (1991).
47. L. Rosenberg, J. R. Palmer, A. G. Zauber, M. E. Warshauer, P. D. Stolley and S. Shapiro, *J. Natl. Cancer. Inst.* **83**, 355-358 (1991).
48. F. M. Giardiello, S. R. Hamilton, A. J. Krush, S. Piantadosi, L. M. Hylind and P. Celano, *N. Engl. J. Med.* **328**, (1993).
49. T. Narisiwa, M. Satoh, M. Sano and T. Takahashi, *Cacinogenesis* **4**, 1225-1227 (1983).
50. D. Kujubu, B. S. Fletcher, B. C. Varnum, R. W. Lim and H. Herschman, *Journal of Biological Chemistry* **266**, 12866-12872 (1991).
51. S. Kargman, G. P. O'Neill, P. J. Vickers, J. F. Evans, J. A. Mancini and S. Jothy, *Proceedings of the American Association for Cancer Research* **36**, 601 (1995).

Biomarkers in Food Chemical Risk Assessment

I. R. McConnell and R. C. Garner

THE JACK BIRCH UNIT FOR ENVIRONMENTAL CARCINOGENESIS, DEPARTMENT OF
BIOLOGY, UNIVERSITY OF YORK, HESLINGTON, YORK YO1 5DD, UK

Abstract

Dietary factors have been implicated in a high proportion of human cancers. These factors
could include direct acting genotoxic agents, compounds which alter the host's response to
endogenous or exogenous genotoxic agents and antioxidant or vitamins. Many of these
determinants can be studied in experimental animals or in humans by the use of
biomarkers. The measurement of DNA adducts is believed to be of particular relevance to
the carcinogenic process. Methods employed for DNA adduct analysis include gas
chromatography-mass spectrometry, immunoassays, physico-chemical methods such as
fluorescence and chromatographic procedures such as ^{32}P-postlabelling. The use of these
techniques and the significance of the results obtained as a measure of target dose and
possibly biological effect with relation to human carcinogen risk assessment are discussed.

Keywords: Molecular epidemiology, dosimetry, DNA adducts, carcinogen risk assessment

Abbreviations: Accelerator mass spectrometry (AMS), adenosine triphosphate (ATP),
aflatoxin B_1 (AFB$_1$), benzo(a)pyrene-7,8-diol-9,10-epoxide (BPDE), bovine serum albumin
(BSA), 7,12-dimethylbenz(a,h)anthracene (DMBA), electron capture detection (ECD), enzyme
linked immunosorbent assay (ELISA), fast atom bombardment (FAB), fluorescence
spectroscopy (FS), gas chromatography (GC), high performance liquid chromatography (hplc),
hepatitis B virus (HBV), ligase mediated polymerase chain reaction (LM-PCR), mass
spectrometry (MS), micrococcal nuclease (MN), nuclease P1 (NP1), negative ion chemical
ionisation mass spectrometry (NCI-MS) polycyclic aromatic hydrocarbons (PAHs),
polyethylene imine (PEI), radioimmuno assay (RIA), spleen phosphodiesterase (SPD),
synchronous fluorescence spectroscopy (SFS), thin layer chromatography (tlc), T4
polynucleotide kinase (T4 PNK), ultra sensitive radioimmunoassay (USERIA).

1 INTRODUCTION

The classical epidemiology studies of Doll and Peto in the UK and Wynder and Gori in the
USA have highlighted the significance of diet in determining cancer incidence[1,2]. Their
investigations indicated that as much as 35% of all human cancer has a dietary component[3,4].
These findings have been further substantiated by the results of detailed studies of migrant
populations and groups from differing socio-economic and religious backgrounds with widely
varying diets and lifestyles[5-7]. However, chiefly due to the complexity of these investigations,

they have largely failed to yield any definitive evidence for a relationship between specific dietary components and cancer aetiology. Correlations have been demonstrated between high dietary fat consumption and cancers of the colon/rectum, gall bladder, prostate, ovary, breast and endometrium[3]. In addition, other dietary factors have been associated with cancers of the liver, stomach, cervix, pancreas, oesophagus, mouth and pharynx[8-10].

Fungal origin

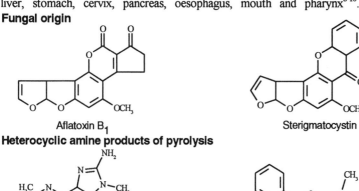

Aflatoxin B$_1$ Sterigmatocystin

Heterocyclic amine products of pyrolysis

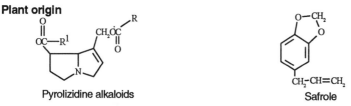

2-amino-3,8-dimethylimidazo(4,5-*f*)quinoxaline   2- amino-1-methyl-6-phenylimidazo(4,5-b)pyridine
(MeIQx)   (PhIP)

Plant origin

Pyrolizidine alkaloids   Safrole

Polycyclic aromatic hydrocarbon products of pyrolysis

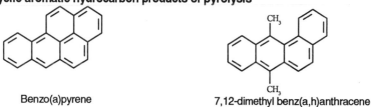

Benzo(a)pyrene   7,12-dimethyl benz(a,h)anthracene

Figure 1   The structures of several recognised and suspect dietary carcinogens

Epidemiological and laboratory animal studies have demonstrated that diet may influence cancer incidence in many different ways, for a comprehensive review see Rogers *et al.*[10] Several recognised genotoxic carcinogens have been detected in foodstuffs or are known to be ingested. These include naturally occurring compounds such as the plant toxin safrole[11] ; carcinogenic secondary fungal metabolites such as aflatoxin B$_1$ (AFB$_1$) produced following colonisation of foodstuffs by some members of the fungal genus *Aspergillus*[12,13]; heterocyclic amines formed as pyrolysis products during the cooking of meat and fish[14,15] ; and polycyclic aromatic hydrocarbons (PAHs) formed as pyrolysis products of cooking or even ingested through the use of oral tobacco[16,17] (see figure 1). Moreover, ingested substances may alter the formation, distribution, activation and detoxification of carcinogens from other sources or affect the promotion of already initiated cell clones and therefore determine the rate of tumour development[18]. Considered of particular importance are the specific effects of total calorific intake[19], dietary fats[20], dietary fibre and micro nutrients such as vitamins A,C,E, carotenoids and selenium, several of which demonstrate important chemoprotective and antioxidant properties[3,8,-10,21].

In order to understand the role of diet in human cancer, fundamental mechanistic studies are required to provide important information concerning the interactions between dietary constituents and the genetic components of the cell. One area of study which will permit more informed risk assessments to be conducted is that of molecular cancer epidemiology and the measurement of biomarkers of carcinogen exposure[22].

1.1 Carcinogen Risk Assessment.

By identifying risk factors for carcinogenesis it is possible to eliminate or reduce exposure by the implementation of suitable control measures, or perhaps the introduction of effective chemopreventative regimens. Risk assessment is defined as the prediction of the probability and nature of the likely consequences of exposure to a chemical carcinogen[23]. Information from studies of structure-activity relationships, short term genotoxicity assays, animal studies and clinical and epidemiological investigations have all been employed for the carcinogenic evaluation of chemicals including dietary constituents [24-26].

1.1.1   Short term genotoxicity assays.   For the purposes of risk evaluation, short term genotoxicity tests provide only a general indication of the potential carcinogenicity of a chemical. Paradoxically, they suffer from the limitation that the end point of the assay is not cancer and that many of them are vastly simplified *in vitro* tests[27]. As a consequence of their simplicity, there are several examples of results which fail to correlate with the *in vivo* situation[28,29].

1.1.2   Experimental animal studies.   Considerable effort has been expended into the development and characterisation of experimental animal models, especially rodents, for the ultimate purpose of controlling human exposure to carcinogens. Studies, although expensive and time consuming, are considered of greater relevance to the human situation than short term genotoxicity tests and are currently the most definitive method available to demonstrate the potential carcinogenicity of a chemical in man[27]. Their use is further substantiated by the observation that all chemicals so far demonstrated as carcinogenic in humans have been shown to induce cancer in one or more animal species[30]. However, there are several features of animal experiments such as the route of administration, the dose administered and the fact that rodents have a much shorter life span than humans which serve to confound the extrapolation of the results of animal studies to man[31,32].

Animal experiments have shown that it is possible to determine a dose level of carcinogen which will eventually lead to tumour formation. These doses are generally high in comparison to those received by humans from environmental exposure [33,34]. Thus, it is also necessary to extrapolate the risk associated from high doses to the low dose levels encountered by man within the environment[35,36,37]. Several complex mathematical models have been developed for this purpose but these generally result in the setting of conservative exposure limits in order to ensure safety[26,38]. Furthermore, it has been concluded that the extrapolation of high doses employed in animal studies to low doses in humans on the basis of mathematical formulae is unsatisfactory and fails to account for essential biological factors such as species differences and human variability in response to carcinogens[33,36,37]. Increasingly, it has become evident that an improved understanding of the fundamental mechanisms of chemical carcinogenesis is required in order to understand inter-species variations in carcinogenic response, and therefore provide data for extrapolation.

1.1.3   Epidemiology investigations.   Epidemiological   studies,   in   principle, provide the ultimate conclusive assessment of carcinogenicity in man. Epidemiology is a direct approach which by-passes the need for extrapolation from animal data[39]. However, as a tool for cancer detection it suffers from a lack of sensitivity and is not generally applicable to explain the general level and wide range of cancers observed in heterogeneous populations. In addition, there are considerable methodological problems in conducting epidemiological studies where multiple comparisons and the influence of confounding factors can lead to equivocal results. They are generally over dependent upon the need for correct information from censuses or self reported exposure history [22,40,41]. For example, there are limitations to the self assessment and monitoring of dietary practice compliance during dietary change studies[8].

Furthermore, epidemiology studies, by their very nature, are retrospective and can only lead to corrective action following the event and manifestation of disease[39,42]. This may take years, during which time the source or extent of exposure may have changed significantly. A major success of epidemiology has been the identification of the link between lung cancer and smoking[43]. Moreover, several of the most notable successes identifying a source of carcinogen exposure have occurred due to perceptive observations by physicians of high incidences of rare cancers within specific groups. These have been largely achieved within the occupational environment where close monitoring of a small group of workers within a controlled environment is achievable[44,45].

Short term genotoxicity assays and animal bioassays generally provide important basic information for assessing risk prior to exposure to most chemicals, however, there is often a need for post exposure surveillance for adverse health effects. Ideally, there should be a more systematic and rapid approach for estimating risk to humans than waiting for epidemiological data which is often of questionable value[22,26,41,46].

1.2 Markers of Exposure to Chemical Carcinogens

It is possible to measure the levels of parent compound in dietary sources using a wide variety of analytical procedures[47]. They provide quantitative estimates of potential intake of a compound or group of compounds which, if excessive may result in the need for elimination of the source of exposure. However, these processes are often compounded by the heterogeneous nature of many environmental contaminants. For example, the distribution of AFB_1 in grain is often uneven due to variations in mould growth throughout a sample[48]. In addition, such measurements fail to account for inter-individual variations in absorption, distribution, metabolism and excretion which may be affected by host factors such as genetic polymorphisms and lifestyle or, moreover, the bioavailability of the carcinogenic material itself[40,49,50,51].

Thus, it has become evident that prospective investigations utilising information from human clinical and highly detailed epidemiological sources are required. These are essential in order to gain important information concerning the doses and mechanisms of action of environmental carcinogens and thereby form more accurate assessments of risk as a result of carcinogen exposure[52]. These so called biological markers may be generally classified into markers of: internal exposure, the level of compound entering the body; biological effect, the biological consequence of exposure; and susceptibility, the likelihood of a deleterious effect[53]. In particular, there has been an enhanced emphasis upon methods aimed at quantifying the extent of exposure to known genotoxins or monitoring individuals for evidence of genotoxic insult in accessible tissues and body fluids[54,55,56].

Internal dose provides a measure of the amount of carcinogen entering the body. Several different markers of internal dose have been employed in biomonitoring studies including direct measurements of the parent compound or a metabolite in body fluids, exhaled air and excreta[57]. Despite providing a less ambiguous measurement of exposure than external dose, this technique fails to give an accurate indication of the concentration of chemical reaching the target site (target dose), which in the case of cancer is DNA. This problem was partially overcome by using techniques directed towards measurement of the biological effect such as cytogenetic analysis of peripheral blood lymphocytes, which provided a measure of toxic response[58,59]. Although of some value, these methods are relatively insensitive and there is uncertainty concerning the phenotypic significance of elevated sister chromatid exchange levels[60]. Quantitative estimates of carcinogen target dose by the measurement of carcinogen macromolecular adducts, however, provides a sensitive measure of the amount of carcinogen interacting at the critical target site and therefore an estimate of the probability of mutation within an individual. This method thereby overcomes differences in species, toxicokinetics, metabolism and detoxification and is therefore more easily equated to the biological effect[46].

1.3 Principles of Biomonitoring for Carcinogen-Macromolecular Adducts

As described above, a preferred method for the analysis/estimation of the amount of chemical or reactive metabolite reaching the critical target site is the formation of macromolecular carcinogen adducts[62]. The interaction of reactive chemicals with a non-target molecule, such as protein, provides a useful means for the estimation of the target dose. Protein adducts, including haemoglobin and albumin, may be readily obtained from blood and have been routinely employed as surrogates for the estimation of target dose of a number of carcinogens including AFB_1 and $PAHs$[63,64]. Adduct formation is proportional to the external dose and, since protein adducts are not repaired, they provide a measure of multiple exposures over a substantial time period, for example the half life of haemoglobin is 100 days[65]. Although protein adducts are not a prerequisite for DNA binding, most DNA adduct forming carcinogens also form blood protein or haemoglobin adducts, which may be analysed with high resolution by gas chromatography-mass spectrometry (GC-MS) or immunoassay[63,66]. However, it must not be assumed that there is a simple relationship between haemoglobin and DNA adducts for all carcinogens[39]. More specifically, DNA adduct formation is related to protein adduct formation and the administered dose only if first order kinetics apply [67].

1.4 DNA Adducts as Markers of Carcinogen Exposure

A causal relationship between electrophilic reactivity and the capacity to induce primary chemical damage to DNA causing mis-replication or erroneous repair leading to mutations and ultimately cancer has been established [68]. Carcinogen induced genetic damage is determined by several factors including exposure, absorption, metabolism and DNA repair (see figure 2)[41]. Each of these may be affected by host factors such as individual dose, genetic polymorphisms and lifestyle [50,51]. DNA adducts formed by reactive carcinogenic species therefore provide an especially useful and direct biomarker for the analysis of target dose[22]. They are an early event in chemical carcinogenesis and yet the end point of exposure, signifying the potential progression to more advanced stages such as mutations within critical genes. Qualitative and quantitative measurements provide pertinent information for the assessment of the biologically effective dose of a carcinogen/mutagen which may then, if possible, be employed to provide an assessment of cancer risk[69].

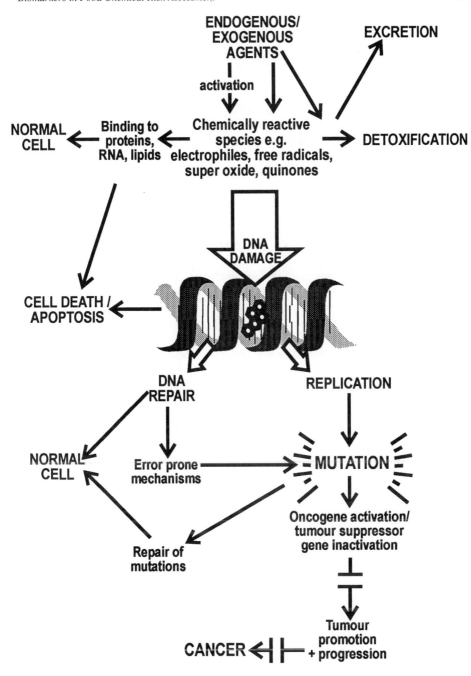

Figure 2
The biological consequences of DNA adduct formation

DNA adduct levels have been studied employing *in vitro* short term test systems and the levels of mutation have generally been shown to correlate well with the levels of mutation observed for critical DNA lesions[70]. In addition, studies using laboratory animals analysing both target and non-target tissues in relation to known tumour responses have led to some understanding of the importance of these adducts in the carcinogenesis process [37,61,71]. Generally, for certain classes of carcinogen including PAH, nitrosamines and mycotoxins, adduct formation may be correlated with carcinogenic potency [61,72,73]. However, this is not always the case. For example highly lipophilic alkylating agents such as ethylene oxide and certain aromatic amines form adducts at similar levels in both target and non target tissues indicating that adduct formation is not the sole reason for carcinogen tissue specificity[73,74]. Thus, although there is no simple rule for the extrapolation of DNA adduct levels in relation to cancer risk within a particular tissue, for certain classes of compound, using animal models, adduct levels have been shown to correlate with tumour yield [37].

2 TECHNIQUES FOR THE MEASUREMENT OF DNA ADDUCTS

The development of carcinogen DNA adduct detection techniques has been rapid in recent times. Much of the impetus stems from the realisation of the potential value of these biochemical markers in chemical carcinogen risk assessment[22,75]. These methods may be employed to detect adducts formed both *in vitro* and *in vivo*. The data can then be used to give insight into tissue distribution, perhaps allowing the identification of target organs or to study the rate of adduct formation, persistence and repair under different experimental conditions[76]. With particular relevance to human biomonitoring, accurate data using highly sensitive techniques are essential in order to measure the target dose. These techniques are a significant development from the information gained from short term genotoxicity assays, animal studies and the retrospective evidence gained from epidemiological investigations.

Technique	Limit of detection femto-moles (1×10^{-15})	Adducts per normal nucleotides	Mass of DNA required (μg)
[32]P Postlabelling			
Standard method	10	1 in 10^7	1-10
Nuclease P1 and butanol enrichment	0.01	1 in 10^{10}	1-10
Immunoassay			
RIA	40	1 in 10^8	up to 10000
ELISA competitive	1	1 in 6×10^8	50
ELISA non-competitive	3	1 in 10^7	0.1
USERIA	1	1 in 10^7	25
Slot-blot	1	1 in 3×10^6	1
Fluorescence			
Low temperature	3	1 in 3×10^8	1000
Synchronous	20	1 in 5×10^8	100
Line Narrowing	1	1 in 10^8	1000
GC-MS	0.5	-	-

Adapted from 63,78

Table 1 The sensitivity of DNA adduct detection methods.

Previously, knowledge of carcinogen metabolism, DNA adduct formation and repair was obtained by employing radiolabelled compounds of high specific activity within *in vitro* and laboratory animal experimental systems. Although still used extensively, the synthesis of radioactive chemicals is very expensive and the location of the label and the stability of the compound may adversely affect the experiment[77]. Moreover, for ethical reasons, it is not possible to study DNA adduct formation in humans by the administration of large amounts of radioactive carcinogen. Consequently much effort has been aimed towards the development of new highly sensitive immunological, biochemical and physico-chemical techniques for the detection of DNA adducts without the need for custom synthesised radioactive compounds[22].

There are several methods available for the detection and quantification of covalent DNA adducts, these include: high performance liquid chromatography (hplc), mass spectrometry (MS), fluorescence spectroscopy (FS), ^{32}P-postlabelling and immuno-assays including enzyme linked immuno-sorbent assay (ELISA), radioimmunoassay (RIA) and ultra sensitive radioimmunoassay (USERIA). Table 1 gives an outline of the current techniques employed for the detection and quantification of covalent DNA adducts and their respective sensitivities, this subject has been extensively reviewed[54,55,63,75,76,78,79]. The ideal characteristics of a method for the detection and quantification of DNA adducts, especially in humans, are: that it is non-invasive, reproducible, inexpensive and can be done quickly with high sensitivity and specificity[22].

2.1 The ^{32}P-Postlabelling technique.

The ^{32}P-postlabelling method is currently the most sensitive technique for the detection and quantification of covalent DNA adducts without the need for custom synthesis of radioactive compounds[79,80]. Devised and developed by Randerath and co-workers, it relies upon the enzymatic transfer of ^{32}P from [γ-^{32}P] ATP to the 5' hydroxyl group of adducted nucleoside 3'-monophosphates by the enzyme T4 polynucleotide kinase (T4 PNK)[81].

Briefly, genomic DNA is digested to deoxyribonucleoside-3'-phosphates using spleen phospodiesterase (SPD) and micrococcal nuclease (MN). The nucleoside 3'-mono-phosphates are then labelled using [γ-^{32}P] ATP and T4 PNK. The resulting radioactively labelled [5'-^{32}P] deoxyribonucleoside 3',5'-bisphosphates are then separated by multi-directional thin layer chromatography (tlc) using a strong anion exchanger such as polyethylene imine (PEI) cellulose, initially to remove normal and then to resolve adducted nucleotides[79,80]. The adduct spots are localised by autoradiography, excised and the radioactivity quantified by scintillation or Cerenkov counting. More recently, quantification has been expedited by the introduction of phosphor-imaging technology[82]. Relative adduct levels (RAL) may then be calculated from the initial quantity of DNA used and the specific activity of the [γ-^{32}P] ATP.

The standard protocol, outlined above allows the detection of approximately 1 adduct in 10^7 normal nucleotides[79-81]. It is most suitable for DNA adducts containing an aromatic or hydrophobic moiety because their chromatographic properties differ sufficiently from normal nucleotides to permit good separation. More recently there have been a series of modifications made to the technique which have considerably enhanced its resolution and more significantly its sensitivity. For example: labelling under ATP limiting conditions where adducted nucleotides are preferentially labelled leading to their intensification[83,84]. However, the most significant advances have been based upon the purification of adducts from normal nucleotides prior to labelling, or the prevention of the latter from being labelled[79]. The most notable and widely used of these are the butanol extraction and the nuclease P1 (NP1) enhancement methods[85,86].

2.1.1   Butanol extraction procedure.   Prior   to   ^{32}P-postlabelling   the digested DNA undergoes a solvent extraction procedure employing butanol containing a phase transfer agent[85]. The relatively hydrophobic aromatic carcinogen adducted nucleotides are extracted into the butanol phase, whilst normal nucleotides and any incomplete digestion products such as dinucleotides remain in the aqueous phase. The adducted nucleotides are then labelled using a molar excess of carrier free [γ-^{32}P] ATP. This extraction procedure serves to increase the sensitivity of the assay to approximately 1 adduct in 10^9 to 10^{10} normal nucleotides, approximating to 1 DNA adduct per cell[80].

2.1.2   Nuclease P1 enhancement method.   The NP1 method developed by Reddy and Randerath has been extensively employed for the analysis of PAH-DNA adducts[87,88]. It involves the further digestion of the digested DNA products with the endonuclease nuclease P1 (NP1) isolated from *Penicillium citrinum*[86]. This enzyme cleaves the 3'-phosphate from normal nucleotides but the presence of some bulky aromatic adducts inhibits its action. T4 PNK requires the presence of a 3'-phosphate group of nucleotides for 5'-phosphorylation and therefore will not label the nucleosides resulting from NP1 digestion. Thus, only the adducted nucleotides form a substrate for ^{32}P-labelling by T4 PNK in the presence of a molar excess of [γ-^{32}P] ATP. This improves the sensitivity to 1 adduct per 10^9-10^{10} normal nucleotides [79].

The ^{32}P-postlabelling technique was a major advance from previous methods of DNA adduct analysis, allowing the measurement of adducts using small amounts of DNA, 1-10 µg, without the need for complex, custom-synthesised radiolabelled compounds[79]. It is not adduct specific and is therefore capable of detecting multiple hydrophobic adducts, following exposure to complex mixtures at sub-femto mole levels. Such quantities are well below the amounts required for characterisation using normal physico-chemical techniques such as mass spectrometry. Thus although co-chromatography with a known standard may provide evidence for the identity of an adduct many of those detected are impossible to characterise fully[80]. Nevertheless, different compounds or groups of compounds yield differing adduct spot profiles and may sometimes be identified from their distinctive patterns, for example the characteristic diagonal radioactive zones observed using NP1 enrichment which are strongly indicative of multiple PAH exposure[87,89]. In addition, the use of this technique has initially succeeded in identifying a novel class of apparently endogenous adduct like, age dependent, DNA binding agents known as I-compounds[90,91].

Despite being a useful tool for the analysis of DNA adducts the ^{32}P-postlabelling methods so far described are only suitable for the quantitative analysis of non-polar, bulky adducts such as polycyclic aromatic hydrocarbons and aromatic amines[79]. Small, relatively polar adducts such as O^6-methyl guanosine may be lost during chromatography due to their structural similarity with normal nucleotides[78,92,93]. Methods have been described to overcome this problem by employing hplc for adduct purification prior to labelling or in order to improve separation of the ^{32}P-labelled adducts afterwards[94,95]. Furthermore, the use of the ^{32}P-postlabelling assay has been confounded by its reliance upon the high fidelity of enzymatic reactions[93,96]. An additional serious drawback to the ^{32}P-postlabelling method is the relatively large amounts of ^{32}P required in order to achieve good quantitative results. [32]P is expensive, laboratories must be well equipped and considerable care exercised to minimise exposure to high levels of radioactivity[78]. Alternative techniques have been tried, such as the use of [γ-^{35}S] ATP and [γ-^{33}P] ATP, which have lower radioactive decay energy and complete avoidance of radioactivity by the use of fluorescent labelling. However, these modifications have generally been accompanied by a corresponding decrease in sensitivity[97-99].

In conclusion, the various modifications of the ^{32}P-postlabelling assay make it a highly sensitive and extremely useful procedure for the analysis of DNA adducts of many types. With continuing developments such as the incorporation of internal standards, purification of substrate DNA and improved adduct resolution, for example by the use of hplc, this technique will continue to provide important information concerning an individual's previous carcinogen exposure[96,100,101].

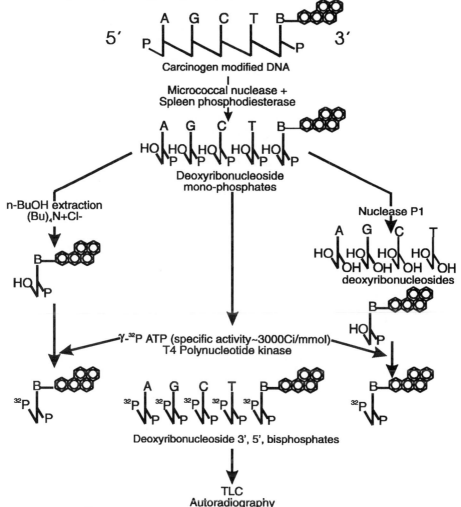

Figure 3 The ^{32}P-postlabelling technique[80].

2.2 Immunoassays

Several strategies and immunisation regimens have succeeded in generating a number of antibodies raised against a diversity of carcinogen modified DNAs and covalently adducted nucleosides[102-104]. In order to produce both polyclonal and monoclonal antibodies a DNA adduct or carcinogen modified DNA is generally linked to a larger protein molecule, such as bovine serum albumin (BSA), in order to facilitate immune recognition. Polyclonal antibody production is relatively simple, requiring the repeated injection of the hapten-protein conjugate into sheep, rabbits or goats for example. The serum is screened for hapten specific antibody which is then purified. In comparison, monoclonal antibody production is laborious and relatively complex. Mice or rats are repeatedly immunised with the hapten protein complex and the serum is screened for specific antibody production. Animals demonstrating a high serum titre are then used for monoclonal antibody production. Briefly, the spleen is removed and antibody secreting splenocytes are immortalised by fusion with an immortal myeloma cell line. The resulting clones are examined for hapten specific antibody production and high serum titre clones are recloned to monoclonality. Finally, monoclonal antibody producing cells are grown using tissue culture techniques or *in vivo* as ascites tumours and the antibody is harvested and purified from the media or ascites fluid.

Most of the antibodies raised against carcinogen adducted DNA, exhibit high affinity and specificity for their respective substrates requiring only microgram quantities of DNA. Adducts at levels as low as 5 adducts per 10^8 normal nucleotides can be analysed, although lower limits are more generally in the region of 1 adduct per 10^7 [103,104]. Both polyclonal and monoclonal antibodies exhibit similar sensitivity and specificity and can be obtained with sufficient avidity to measure carcinogen modification in biological samples[104,105]. Polyclonal antibodies generally require less work, however, monoclonals can be produced from smaller quantities of immunogen and once a clone has been established offer the advantage of unlimited homogeneous supplies.

ANTIGEN	REFERENCE
O^6-Methyl deoxyguanosine	106, 107.
O^2-Methyl deoxythymidine	102
O^6-Ethyl deoxyguanosine	108, 109, 110, 111
O^4-Ethyl deoxythymidine	106, 108
O^6-Butyl deoxyguanosine	106, 108
Acetyl aminofluorene-C8-deoxyguanosine	112, 113, 114, 110
Amino fluorene-C8- deoxyguanosine	115,116
Acetyl amino fluorene adducted DNA	117, 118
Benzo(a)pyrene-7,8-diol-9,10-epoxide-I adducted-DNA	119, 120, 121
Cis platin adducted DNA	122, 123, 124, 125
1-Amino pyrene adducted-DNA	126
8-Methoxy psoralen adducted -DNA	127
Aflatoxin B_1 adducted -DNA	128, 129, 130
Oxidatively damaged DNA containing thymine glycols	131, 132
UV damaged-DNA {Thymidine dimer}	133, 134
Melphalan adducted -DNA	135

Table 2

Examples of antibodies raised against carcinogen-DNA adducts (adapted from 78, 103)

Antibodies may be employed in immunoassays for the detection of DNA adducts including competitive radioimmunoassays and solid phase assays such as enzyme linked immunosorbent assay (ELISA) and ultrasensitive enzyme radio immuno assay (USERIA)[103,104]. For solid phase assays, the modified DNA is bound to a solid surface, treated with the antibody, and a species specific second antibody conjugated to an enzyme or a fluorescent dye is added. Determination of the amount of bound fluorescence or enzyme activity is used to quantify the amount of DNA adduct from a standard curve. As little as 30 fmoles of adduct per mg DNA can be detected using 20 μg DNA, although sensitivity is largely dependent upon the detection system employed[76,78]. The major advantage of the use of competitive solid phase immunoassays is their ease and reproducibility allowing the relatively inexpensive analysis of large numbers of samples making it readily applicable to molecular epidemiological studies and, with most ELISAs, the avoidance of radioactive materials[22,75].

The principal limitation of the antibody approach for the detection of covalent DNA adducts is the need to develop specific antibodies towards each adduct of interest[102]. Considerable time and effort is required to produce and characterise each antibody. Production requires the chemical synthesis of relatively large quantities of nucleotide adducts or highly modified DNA produced by the reaction of DNA with the reactive intermediate of a carcinogen. This DNA is essential not only for immunisation and characterisation but also for use as standards during the assay itself. Furthermore, knowledge of the complete adduct structure assures the antibody is recognising the antigen of interest and not a contaminant. Thus, for the sake of clarity it is often necessary to analyse relatively large amounts of the synthesised adducted sample by physico-chemical methods[78].

In order to elicit a good immune response it is necessary to raise antibodies towards a highly modified chemically synthesised adducted standard DNA sample. Such antibodies may recognise more highly adducted DNA with much higher efficiency than lower levels, resulting in discrepancies in quantification[136,137]. Recognition is also affected by the state in which the DNA is presented to the antibody for example single stranded, partially digested DNA or mononucleotides[103]. In addition, antibodies may cross react with related DNA adducts with varying affinities thereby affecting the detection of specific adducts within complex mixtures[127]. For example, several antibodies raised against benzo(a)pyrene-7,8-diol, 9,10 epoxide DNA have demonstrated significant cross reactivity towards other PAHs such as dimethyl benz(*a*,*h*)anthracene (DMBA)[138]. Cross-reactivity can evidently give rise to problems of specificity, which may need to be overcome by prior separation of adducts. Nevertheless, it may also be turned to advantage in order to assist the characterisation of as yet unrecognised DNA adducts, for example background adducts occurring in human DNA[139].

2.3 Immunocytochemistry

Besides immunoassays, antibodies may also be employed for the analysis of DNA adducts at the cellular level using immunocytochemical techniques[140]. Addition of a second species specific enzyme linked or labelled antibody allows visualisation of the initial antibody adduct interaction which may then be quantified using image analysis techniques. These assays are less sensitive than ELISA but can yield information concerning the distribution of adducts within a cell population, for example aflatoxin adducts in the liver or BPDE adducts in white blood cells[141-143]. Furthermore, antibodies have been employed to study the distribution of adducts at the level of the gene. In combination with electron microscopy, the binding of antibodies to a BPDE modified DNA sequence was visualised and the site of DNA adduct formation localised[144]. Similarly, antibodies raised against BPDE-DNA have been employed as tools for the analysis of the structure and conformation of PAH-diol epoxide DNA adducts[145].

2.4   Immuno-concentration

In addition to the immunological assays outlined above, antibodies have been immobilised onto Sepharose support matrices and employed in the form of an immunoaffinity column[138,145,146]. These provide a useful preparative technique for the concentration or isolation of specific adducts prior to analysis using other methods. This has been used as an enrichment procedure prior to ^{32}P-postlabelling and hplc and offers the advantage of increased sensitivity, specificity and the opportunity to measure different adducts within the same sample[138,147].

2.5   Fluorescence Spectroscopy (FS).

There are a number of chemical carcinogens with rigid, conjugated structures which when activated by light, absorb energy and then re-emit it at a higher, characteristic, wavelength (lower energy) i.e. fluorescence. First used by Miller in 1951 to demonstrate the binding of BaP to mouse skin proteins, fluorescence has become the basis of some of the current techniques used to study the covalent binding of carcinogens to cellular macromolecules[148]. Evidently, fluorescence spectroscopy is limited solely to those carcinogens which undergo fluorescence. However, PAH and aflatoxin adducts, which both display this property, are widely recognised as significant and often widespread carcinogenic hazards and as such are frequent subjects for covalent adduct investigations[76].

Initially the sensitivity of fluorescence techniques was generally not sufficient to detect the low levels of DNA adducts normally found *in vivo*. Several adaptations have been made in order to enhance sensitivity, such as the introduction of fluorescence line narrowing spectrometry, but excitation and emission spectra are often complex with many peaks[78]. The main fluorescence technique to be applied to the analysis of DNA adducts has been Synchronous fluorescence spectroscopy (SFS), often used following hplc purification. SFS spectra are obtained by varying both excitation and emission wavelengths whilst keeping a constant wavelength interval between the two spectra. Hence it is possible to detect and identify fluorescent molecules from their highly characteristic excitation and emission spectra such that SFS has the potential for ultimately detecting individual components present within complex mixtures. So far it has been employed principally for the analysis of benzo(*a*)pyrene and AFB$_1$-DNA adducts, and to a far lesser extent for methyl and etheno adducts[149-151]. The lowest reported detection limit, obtained employing BPDE standards, is currently 3 adducts in 10^8 nucleotides using 100 µg DNA which is less sensitive than ^{32}P-postlabelling yet comparable with immunoassays[152]. However, SFS offers the significant advantage that the DNA samples are not destroyed during analysis allowing further studies using alternative techniques[153].

2.6   Gas Chromatography and Mass Spectrometry

Mass spectrometry (MS) is a definitive technique for the analysis of trace organic compounds. Gas chromatography (GC) provides excellent separation which when coupled to a highly sensitive electron capture detector (ECD) or a negative ion chemical ionisation mass spectrometer (NCI-MS) provides a powerful analytical tool. Gas chromatography and gas chromatography-mass spectrometry (GC-MS) techniques have now been developed with sufficient sensitivity to allow the analysis of biological samples[75,78]. In particular good correlations have been observed for exposure to several carcinogens and the formation of haemoglobin adducts [154,155].

Although more extensively employed for the analysis of protein adducts, GC-ECD and GC-NCI-MS are increasingly being applied to DNA[156,157]. The procedures used to isolate the adducts from the DNA are fundamental to the sensitivity of the assay but generally they require hydrolysis of DNA to liberate the base adducts from the nucleoside. For higher sensitivity it is also necessary to derivatise the DNA adducts to produce an electrophore facilitating detection down to femto mole concentrations[157].

Much of the application of GC-MS to DNA adduct analysis has centred upon the detection of adducts *in vitro* or those resulting from the administration of carcinogens to laboratory animals. Studies of note include the analysis of 7-methyl guanine adducts in rats administered three potential alkylating agents and the analysis of oxidative base damage[154,158]. There have been few studies using GC-MS for the analysis of human samples. Adducts analysed to date include BPDE, tobacco specific nitrosamine malondialdehyde and 7-methyl ring opened guanine DNA adducts, although the true significance of the latter measurements is equivocal due to the presence of high background levels[159,160,161]. These techniques have potential for high sensitivity and high resolution, although further work is required before they can be routinely used for the analysis of DNA adducts from human samples. The role of GC-MS, in particular, is likely to be promoted by improved strategies and advances in sensitivity caused by the increasing use of soft ionisation techniques such as Fast Atom Bombardment mass spectrometry (FAB-MS), negative electrospray ionisation mass spectrometry and matrix - assisted laser desorption mass spectrometry[157,162,163]. In addition, the application of tandem MS machines where a fragment generated in the first machine may be further selectively fragmented in the second may provide a means of identifying adducts in mixtures[164]. Furthermore, Accelerator Mass Spectrometry (AMS), theoretically capable of detecting zeptomole levels $(1 \times 10^{-21}$ moles) or 1 adduct/10^{12} nucleotides of ^{14}C labelled DNA adducts, although requiring the use of very low levels of radioactivity, could provide important evidence concerning the toxicokinetics of adduct formation and repair as well as the biological consequences of low levels of adducts in humans[165,166]. Nevertheless, time and cost factors are considerable for studies involving MS and consequently the main use for such methodologies is currently in the confirmation and in-depth qualitative analysis of samples shown to contain DNA adducts by more rapid, less intensive means[78].

2.7 Molecular Biology

The increased availability of cloned DNA modifying enzymes has led to the development of several molecular biology based assays for the detection and quantification of DNA lesions. Souliotis *et al.* employed the DNA repair enzyme O^6-alkyl transferase in a competition assay for the measurement of alkyl adducts[167]. This procedure used an endlabelled synthetic oligonucleotide standard containing a single O^6-methyl guanine adduct which was incubated with the test DNA in the presence of O^6-alkyl transferase. The extent of synthetic oligonucleotide repair was inversely proportional to the adduction level of the test DNA used. Therefore, the level of adduction could be quantified from the extent of repair of the standard O^6-alkyl guanine which was measured following immunoprecipitation of unrepaired oligonucleotide using antiserum raised against O^6-alkyl deoxyguanosine [167]. Detection levels were estimated at 1 methyl guanine adduct in 10^8

normal nucleotides with the sensitivity of ethylated guanines being slightly lower. This technique has been successfully employed for the quantification of methyl adducts resulting from exposure to the chemotherapeutic agent procarbazine in human blood and alkyl adducts in normal and atrophic gastric mucosa[168,169].

Recent evidence has shown that DNA repair and mutations within critical target genes are fundamental to the cancer process[170-174]. This has led to the development of methods for the analysis of DNA damage and repair within specific genes. Although in relatively early stages of development and yet to be successfully applied to human biomonitoring, several of these techniques demonstrate good prospects for the future. Initial studies by Bohr *et al.* applied a Southern blot adaptation originally devised by Nose and Nikaido[175,176]. The adducted DNA was partially digested, treated with the pyrimidine dimer specific T4 endonuclease V enzyme and probed for specific sequences by Southern hybridisation in order to detect the positions of adduction. The presence of an adduct was indicated by cutting and eventual disappearance of a fragment from the Southern blot. By replacing the T4 endonuclease V with the nucleotide excision repair system the (A)BC excinuclease a general method for the quantification of DNA adducts in defined sequences of mammalian genomic DNA was devised[177]. However, this technique required relatively high concentrations of DNA and a specific restriction site adjacent to the gene of interest and was not sufficiently sensitive for the detection of adducts at low levels. In addition, for strand specific repair studies at non lethal adduct densities a lower limit of resolution of only several kilobases was possible[178].

More recently polymerase chain reaction (PCR) technology has been applied towards the detection of DNA lesions. One such example has been the application of a quantitative PCR methodology[179]. The speed and ease of PCR coupled with the need for only small amounts of template and no specific endonucleases make it readily applicable to use in molecular epidemiological techniques. *Taq* polymerase is inhibited by many DNA lesions thereby causing a decrease in the amplification of damaged relative to non damaged DNA. Poisson analysis may then be employed in order to quantify the lesion frequency. For cisplatin adducts, this technique was shown to be 3-5 fold more sensitive than the Southern blot (A)BC excinuclease assay described above[177]. However, not every DNA lesion is capable of blocking *Taq* polymerase and quantitative PCR techniques can be susceptible to variations and duplication of results can be difficult.

A further application of PCR has been the use of ligation mediated PCR (LMPCR) for the detection of DNA adducts. LMPCR is a single directional PCR in which the sequence of only one end of the desired sequence is required for amplification[180]. The sequence position of adducts can be mapped from as few as 100 molecules of DNA whenever it is possible to convert the adduct chemically or enzymatically into a strand break with a 5'-phosphate group. This has been carried out chemically employing piperidine to cleave alkali labile sites and enzymatically employing T4 endonuclease V/DNA photolyase and the (A)BC excinuclease[178]. Gene specific oligonucleotide primers are subsequently employed to selectively amplify and detect the sequences of interest[178]. This technique has been employed for the detection of cyclobutane dimers, 6-4 photoproducts and alkylation sites within specific genes and oncogenes[181-183]. The ability of LMPCR to map low frequency adducts along single copy genes is mainly limited by the specificity of the adduct cleavage system. It has been proposed that future work will apply this procedure to several other adducts employing DNA glycosylases including FAPY and 3-methyl adenine glycosylases which are known to cleave DNA at specific adducted deoxynucleotides.

2.7 Summary

Methods for the detection of carcinogen macromolecular adducts and in particular DNA adducts, provide the means not only to monitor exposure to recognised carcinogenic sources but also unidentified ones. In order to prove useful for biomonitoring studies a level of sensitivity of 1 adduct in 10^8 normal nucleotides is necessary although the amount of DNA required is equally significant. These adduct levels generally correspond to femtomole and attomole levels in micrograms of DNA. ^{32}P-postlabelling, SFS and immunoassays such as ELISA are sufficiently sensitive and are widely employed for the quantification of DNA adducts in human tissues. Each has specific advantages and disadvantages but it is important to recognise that data obtained in employing one biomonitoring technique is considerably less reliable than that substantiated by duplication using an alternative system [42,76].

3 THE MEASUREMENT OF DNA ADDUCTS IN HUMANS

Humans are generally exposed to a wide diversity of carcinogenic substances. The methods outlined above have been employed for the analysis of tissues obtained during surgery or more often following autopsy. These analyses have succeeded in demonstrating human exposure to a wide range of chemical carcinogens including PAHs and AFB_1[87,88,143]. However, true biomonitoring studies are limited by the fact that the target tissue is often not available for the analysis of DNA adducts. For this reason, human biomonitoring is generally limited to the analysis of surrogate tissues or excreta obtained by less invasive means. The most common surrogate tissue to be used is white blood cells or lymphocytes, others include oral mucosa and placenta[184-186]. However, adducts detected by non invasive means are not always derived from the target tissue and thus the relationship between adducts in non target tissues and the target dose may be complex as has been demonstrated[187].

Recently several extensive investigations have shown the potential value of human biomonitoring for DNA damage. Perera *et al.* employed a battery of six biological markers to measure molecular and genetic damage in peripheral blood from residents of a highly industrialised and environmentally polluted region of Poland with respect to residents of a rural less polluted area over a 10 month period. There were significant differences in the levels of PAH and other aromatic carcinogen DNA adducts with winter levels exceeding summer and industrial exceeding rural which correlated with the levels of atmospheric air pollution. In addition, environmental pollution was associated with significant increases in chromosomal aberrations, sister chromatid exchange and the frequency of increased *ras* oncogene expression. Furthermore, the presence of PAH-DNA adducts exhibited a significant correlation with chromosomal mutation. This evidence suggested a possible link between environmental exposure to PAH and genetic alterations relevant to disease[188].

The most definitive proof of an association between exposure and disease outcome is demonstrated by prospective molecular epidemiological studies where healthy individuals are monitored until the diagnosis of disease[48]. Although the use of nucleated blood cells for the analysis of carcinogen exposure provides a more practicable and less invasive means of biomonitoring carcinogen exposure than collecting tissues during surgery, it remains a tedious process which in many cases serves to limit the number of samples that can be taken. In contrast to monitoring DNA adducts *in situ*, the eliminated products of carcinogen DNA interactions can be monitored easily and non-invasively by the analysis of urine[189]. Urine collection is easy and it is relatively safe to handle compared to other bodily fluids. A recent study has analysed the urinary excretion of AFB$_1$-guanine in the Guangxi autonomous province of China where there is a high incidence of liver cancer. Urinary AFB$_1$-guanine excretion was analysed by immuno-concentration followed by hplc and was shown to correlate with AFB$_1$ intake[190]. This marker had previously been shown to be a good marker of hepato-carcinogenic risk in animal models[191]. These methods were also employed in an extensive prospective study into the relationship between markers of AFB$_1$ and HBV and the development of liver cancer in Shanghai province, China. Over a 7 year period, urine samples were collected from males between the ages of 45 and 64 and analysed for the presence of AFB$_1$ biomarkers, including AFB$_1$-guanine, and HBV surface antigen. During the course of the study several developed liver cancer. A highly significant increased risk of developing liver cancer was observed for individuals where AFB$_1$-guanine was detected in the urine and those demonstrating the HBV antigen. Furthermore, those individuals found to be exposed to AFB$_1$ and HBV positive had an increased chance of developing liver cancer. These results demonstrated a relationship between the presence of carcinogen biomarkers and cancer risk and moreover, a synergistic interaction between AFB$_1$ and HBV both of which are recognised risk factors for hepatocarcinogenesis [192].

4   DOSIMETRY AND RISK ASSESSMENT

The measurement of DNA adducts provides the only assured method for establishing whether specific chemicals are capable of interacting with the host cell genome following evasion of cellular detoxification pathways. DNA adduct formation and mutation are incontrovertibly linked. Furthermore, evidence strongly suggests that cancer is a mutational process. Thus, DNA adduct measurement in addition to providing an estimate of internal dose may provide an indication of cancer potential. Molecular epidemiology has demonstrated that carcinogen exposure in humans results in similar genotoxic damage to carcinogen dosed experimental animals[170,193]. So it appears that animal models do have some relevance to man and therefore in risk assessment. Most human exposure is assumed to be continuous. Thus it is presumed that adduct measurements represent a steady state reflecting recent carcinogen exposure, metabolism, tissue turnover and DNA repair[76]. Adduct levels are presently assessed largely with the aim of identifying marked increases in susceptibility due to genetic polymorphism, for instance, or to become aware of previously unrecognised exposures. In addition, they may provide information concerning the distribution, rate of formation, persistence, and repair of adducts in target and non-target tissues or relative to tissues of non exposed individuals. Little is known about the long term biological effects of the low levels of carcinogen DNA adducts normally found in human tissues[62]. Inter-species extrapolations, even when based on measurements of the biologically effective dose, must be interpreted with caution since low DNA adduct levels are not necessarily indicative of low tumourigenic risk[37]. How these genetic changes relate to tumour development in humans is yet to be established. Thus, to facilitate risk assessment there is a need to relate DNA adducts to genetic changes.

Informed estimates of risk may be made on the basis of model systems such as the parallelogram approach pioneered by Fritz Sobels[194]. A specific biomarker in an experimental system is employed and associated with the biological outcome in that experimental system, namely cancer. The same biomarker is then measured in the equivalent human tissue and this value compared with that found in the experimental system. Finally, the data obtained from the human biomarker such as adduct concentration is extrapolated in order to obtain an ultimate cancer risk.

One further aid in the risk assessment process is the use of a 'rad-equivalence' approach pioneered by Ehrenberg and co-workers[195,196]. This theory applies the fact that the corresponding ultimate deleterious biological effect of exposure to genotoxic carcinogens and radiation is the induction of mutations in DNA. Differences in the induction of mutations between species result largely from species specific factors such as the relative efficiencies of repair and the frequency of cell replication. By obtaining values for the induction of mutations in different species exposed to the same target dose of a genotoxic chemical it is possible to compensate or account for such variation and therefore provide a means of prospective risk assessment by improving the accuracy of extrapolation of data from experimental models to man. Furthermore, a major advantage is the ability to identify and estimate low risks of cancer and the use of a common unit (rad-equivalents) which facilitates comparisons of the promutagenic capacity of adducts and moreover, the risks posed from exposure to different compounds. Central to both the radiation dose equivalence and parallelogram concepts is the need for an accurate determination of the target dose which is most accurately assessed by the precise measurement of DNA adducts. Whilst these approaches exhibit inadequacies such as failing to account for species differences in metabolism and repair they are far superior to the methods of high dose/low dose extrapolation from animal studies which are currently employed for the setting of regulatory limits.

Increasingly it has become recognised that DNA damage alone is not sufficient for the induction of cancer[170]. Chemical carcinogenesis is a complex process dependent upon numerous exogenous and endogenous factors. It is now strongly believed that mutations *per se* may not result in cancer induction. Moreover, the accumulation of mutations within oncogenes and tumour suppressor genes is believed to ultimately result in malignant transformation. Evidence has also shown that the rate of DNA repair may differ between genes and even transcribed and non transcribed strands [197,198]. This has important implications for genetic toxicology because mutations induced by carcinogens occur preferentially in the non transcribed strand. The position of adduct formation within the genome is therefore fundamental and may serve to question the value to risk assessment of results obtained employing whole genome techniques such as ^{32}P-postlabelling and ELISA.

CONCLUSIONS

Human exposure comprises multiple chemical carcinogens. Genotoxic carcinogens within the diet are a recognised widespread hazard especially within some areas experiencing hot, humid climates due to the poor storage of foodstuffs. Moreover, the increasing passage of foods throughout the world may lead to exposure within developed countries unless stringent regulations to ensure the standards of foodstuffs are enforced. Furthermore, within the Western world the thorough cooking of foods, especially meat and fish products for example, in order to avoid bacterial food poisoning may result in a paradoxical increase in the frequency and quantity of ingestion of dietary carcinogens formed as pyrolysis products.

Studies into human DNA adduct measurement remain at a relatively early stage. However, they currently provide important information concerning an individual's previous carcinogen exposure and in some instances their ultimate susceptibility to cancer induction. The quantification of DNA adducts at the extremely low levels found following low dose carcinogen exposure requires the continued development of highly sensitive analytical techniques. In addition, techniques must be capable of distinguishing between

DNA adducts of different types and different adducts of the same carcinogen with differing promutagenic properties. In the absence of preparative procedures, several of the techniques outlined above are generally unable to provide accurate measurements of specific adducts or sufficient sample for chemical characterisation. Recently, combinations of methods such as immuno-affinity concentration prior to analysis by hplc, ^{32}P-postlabelling or mass spectrometry have allowed the identification of specific adducts in human tissues and are likely to lead to the identification of new as yet unrecognised sources of exposure. Currently, the contrasting sensitivities of adduct detection using the ^{32}P-postlabelling technique and physico-chemical characterisation is a major obstacle to identification[139]. It is believed that advances within the field of mass spectrometry in particular will have an important role in the chemical characterisation of adducts so far not characterised. However, it is fundamental that newly developed assays are fully characterised using standards and laboratory animals prior to their employment in human studies.

Adduct measurements are generally made at a particular time point and therefore only represent a transient observation. The number of lesions detected over a longer time course may provide a more comprehensive measure of adduct formation and repair. In addition, investigations are required into the role of endogenous DNA lesions. Above all, it is fundamental to establish the relevance of particular biomarkers within the cancer process if they are to provide pertinent information concerning the cancer outcome and not simply a measure of internal dose. Ultimately it is hoped that the measurement of biomarkers will provide a reliable assessment of risk which may be employed to identify individuals likely to develop cancer and therefore allow intervention studies to be undertaken such as the prophylactic administration of antioxidants[48].

ACKNOWLEDGEMENTS

IRM is grateful for financial support from Zeneca Pharmaceuticals. RCG is supported by several funding bodies including MAFF, York Against Cancer and the European Commission as well as industrial companies such as Zeneca Pharmaceuticals, Glaxo, SmithKline-Beecham, Astra, Bayer, Pharmacia and Shell. We are grateful to them for their support.

REFERENCES

1.   R. Doll and R. Peto, *Journal of the National Cancer Institute.*, 1981, **66**, 1193.
2.   E. L. Wynder and G. B. Gori, *Journal of the National Cancer Institute,* 1977, **58**, 825.
3.   R. Doll *Cancer Res.*, 1992, **52(SUPPL.)**, 2024s.
4.   E. L. Wynder, *Archives of Surgery*, 1990, **125**, 163-169.
5.   M. Thorogood, J. Mann, P. Appleby, K. McPherson, *British Medical Journal*, 1994 **308**, 1667.
6.   J.L. Lyon, M.R. Klauber, J.W. Gardner, C.R. Smart, *New England Journal of Medicine*, 1976, **294**, 129.
7.   R.L. Phillips and D.A Snowdon, *Journal of the National Cancer Institute*, 1985, **74**, 307.
8.   P. Greenwald and C. Clifford *in Cancer Prevention and Control*, eds., P. Greenwald, B.S. Kramer, D.L. Weed, Marcel Dekker, New York 1995, pp 303.
9.   B.S. Reddy, L.A. Cohen, G.D. McCoy, P. Hill, E.L. Weisburger, E.L Wynder *Advances in Cancer Research*, 1980, **32,** 237.
10   A.E.Rogers, S.H. Zeisel, and J. Groopman, *Carcinogenesis*, 1993, **14**, 2205.

11   J.A. Miller, A.B. Swanson, E.C. Miller, in *Naturally occurring carcinogens-mutagens and modulators of carcinogenesis* eds.E.C. Miller, J.A. Miller, I. Hirono, T. Sugimura and S.Takayama, Japan Scientific societies press, Baltimore, 1979, pp 111.

12.   I.R. McConnell and R.C. Garner, *in DNA adducts: Identification and Biological Significance* eds., K. Hemminki, A. Dipple, D.E.G. Shuker, F.F. Kadlubar, D. Segerbäck and H. Bartsch, IARC Scientific Publications No. 125 Lyon 1994, pp 49.

13.   R.C. Garner, M.M. Whattam, P.J.L. Taylor and M.W. Stow, *J. Chromatogr.*, 1993, **648**, 485.

14.   B. Wakabayashi, M. Nagao, H. Esumi and T. Sugimura, *Cancer Res.*, 1992, **52 (SUPPL.)**, 2092.

15.   D.W.Layton, K.T. Bogen, M.G. Knize, F.T. Hatch, V.M. Johnson and J.S. Felton, *Carcinogenesis*, 1995, **16**, 39.

16.   Lijinsky, *Mutat. Res.*, 1991, **259**, 251.

17.   G. Grimmer, in *Environmental carcinogens selected methods of analysis* eds., H. Egan, M. Castegnaro, H. Kunte, P. Bogovski, E.A. Walker, W. Davis, Vol. 2, IARC Scientific Publications No. 29, Lyon, 1979, pp 31.

18.   A.H. Conney, in *Naturally occurring carcinogens-mutagens and modulators of carcinogenesis* eds.E.C. Miller, J.A. Miller, I. Hirono, T. Sugimura and S. Takayama, Japan Scientific societies press, Baltimore, 1979, pp 139.

19.   W.K. Lutz and J. Schlatter, *Carcinogenesis*, 1992, **13**, 2211.

20.   L.A. Cohen, *Scientific American*, 1987, **257**, 42.

21.   B.N. Ames, *Science* 1983, **221**, 1256.

22.   R.C. Garner, *Carcinogenesis*, 1985, **6**, 1071.

23.   I.F.H. Purchase, *Human Toxicology*, 1990, 8, 175.

24.   J. Ashby and R.W. Tennant, *Mutat. Res.*, 1988, **204**, 17.

25.   T.H. Maugh, *Science*, 1987, **202**, 37.

26.   W.P. Watson, C. Bleasdale and B.T. Golding, *Chem. Brit.*, 1994, **30**, 661.

27.   G.R. Douglas, D.H. Blakey and D.B. Clayson, *Mutat. Res.*, 1988, **196**, 83.

28.   B.A. Bridges, *Mutat. Res.*, 1988, **205**, 25.

29.   J. Ashby and R.S. Morrod, *Nature*, 1991, **352**, 185.

30.   *IARC Monographs on the Evaluation of the Carcinogenic risk of Chemicals to Humans*, Supplement 7. International Agency for Research on Cancer, Lyon, 1987.

31.   I. F. H. Purchase, in *Cancer Risks: strategies for elimination*, ed. P. Bannasch, Springer-Verlag, Berlin, 1986, pp. 67.

32.   J. Huff, J. Haseman and D. Rall, *Annual Review of Pharmacology and Toxicology*, 1991, **31**, 621.

33.   S.M. Cohen and L.B. Ellwein, *Chemical Research in toxicology.*, 1992, **5**, 742.

34.   L.J. Marnett, *Chemical Research in toxicology.*, 1992, **5**, 735.

35.   L.S. Gold, T.H. Slone, B.R. Stern, N.B. Manley and B.N. Ames, *Science* 1992, **258**, 261.

36.   B.N. Ames and L.S. Gold, *Proc. Natl. Acad. Sci. USA*, 1990, **87**, 7772.

37.   M.C. Poirier and F.A. Beland, *Chemical Research in toxicology.*, 1992, **5**, 749.

38.   W.K. Lutz, *Carcinogenesis*, 1990, **11**, 1243.

39.   J. Ashby, *Mutat. Res.*, 1988, **204**, 543.

40.   C.C. Harris, *Environmental Health Perspectives.*, 1985, **62**, 185.

41.   C.C. Harris, *Cancer Res.*, 1991, **52 (SUPPL.)**, 5023s

42.   P.G. Shields and C.C. Harris, *Journal of the American Medical Association.*, 1991, **266**, 681.

43.   R. Doll and A.B. Hill, *British Medical Journal*, 1952, **2**, 1271.

44.   L. Rehn, *Arch. Klin. Chir.*, 1895, **50**, 588.

45.   J.L. Creech and M.N. Johnson, *Journal of Occupational Medicine*, 1974, **16**, 150.

46.   Aaron, in *Genetic Risk Assessment Monograph 1 of the Environmental Health Institute* ed. A.D. Bloom and P.K.F. Poskitt, March of Dimes, New York, 1988, pp53.

47. *Environmental carcinogens selected methods of analysis* eds., H. Egan,M. Castegnaro, H. Kunte, P. Bogovski, E.A. Walker, W. Davis, Vol. 1 and 2, IARC Scientific Publications No. 29, Lyon, 1979.
48. J.D. Groopman and T.W. Kensler, *Chemical Research in toxicology.*, 1993, **6**, 764.
49. G.N. Wogan,*Environmental Health Perspectives.*, 1989, **81**, 9.
50. J.R. Idle, *Mutat. Res.*,1991, **247**, 259.
51. D.W. Nebert, *Mutat. Res.*, 1991, **247**, 267.
52. W.K. Lutz, J. Poetzsch, J. Schlatter and C. Schlatter, *Trends in Pharmacological Science*, 1991, **12**, 214.
53. J.D. Groopman, W.D. Roebuck, T.W. Kensler, *in Diet and cancer: markers, prevention and treatment* ed M.M. Jacobs, Plenum press New York, 1994,pp 149.
54. R.A. Santella, *Mutat. Res.*, 1988, **205**, 271.
55. H. A. J. Schut and K. T. Shiverick, *FASEB J.*, 1992, **6**, 2942.
56. D. E. G. Shuker and P. B. Farmer, *Chemical Research in toxicology.*, 1992, **5**, 450.
57. M.E. Andersen, *Cell Biology and Toxicology*, 1989, **5**, 405.
58. P.E. Perry and H.J. Evans, *Nature*, 1975, **258**, 121.
59. M. Sorsa, K. Hemminki and H. Vainio, *Teratogenesis, Carcinogenesis and Mutagenesis*, 1982, **2**, 137.
60. Soper, P.D. Stolley, S.M. Galloway, J.G. Smith, W.W. Nichols and S.R. Wolman, *Mutat. Res.*, 1984, **129**, 77.
61. W.K. Lutz, *Mutat. Res.*, 1979, **65**, 289.
62. F. P. Perera, *Mutat. Res.*, 1988, **205**, 255.
63. P.B. Farmer, H.-G. Neumann and D. Henschler, *Arch. Toxicol.*, 1987 **60**, 251.
64. F. P. Perera, J. Mayer, R.M. Santella, D. Brenner, A. Jeffrey, L. Latriano, S. Smith, D. Warburton, T.L. Young, W.Y. Tsai, K. Hemminki, and P. Brandt-Rauf *Environmental Health Perspectives.*, 1991, **90**, 247.
65. S.Osterman-Golkar, L. Ehrenberg, D. Segerbäck, and I. Hallstrom, *Mutat. Res.*, 1976,**34**, 1.
66. C. Wild, Y.Z. Jiang, G. Sabbioni, B. Chapot and R. Montesano, *Cancer Res.*, 1990, **50**, 245.
67. L. Ehrenberg and S. Osterman-Golkar, *Teratogenesis, Carcinogenesis and Mutagenesis* 1980, **1**, 105.
68. K. Hemminki, *Carcinogenesis*, 1993, **14**, 2007.
69. F.P. Perera and P. Boffetta, *Journal of the Naional Cancer Institute.*, 1988, **80**, 1282.
70. D. Lawley, *Mutat. Res.*, 1974, **23**, 283.
71. P. Perera, M.C. Poirier, S.H. Yuspa, S. Nakayama, A. Jaretzki, M.M. Curnen, D.M. Knowles and I.B. Weinstein, *Carcinogenesis*, 1982, **3**, 1405.
72. S.W. Ashurst, G.M. Cohen, S. Nesnow, J. Di Giovanni and T.J. Slaga, *Cancer Res.*, 1983, **43**, 1024.
73. D. Segerback, *Chemico.-Biological Interactions* 1983, **45**, 139.
74. N-G. Neumann, *Journal of Cancer Researchand Clinical. Oncology*, 1986, **112**, 100.
75. P.T. Strickland, M.N. Routledge and A. Dipple, *Cancer Epidemiology, Biomarkers & Prevention*, 1993, **2**, 607.
76. A. Weston, D.K. Manchester, A. Povey and C.C. Harris, *Journal of the American College of Toxicology*, 1989, **8**, 913.
77. Baird, in *Chemical carcinogens and DNA* ed. P.L. Grover, CRC Press, Boca Raton, 1979, pp. 59.
78. D.H. Phillips, *in Chemical Carcinogenesis and Mutagenesis* Vol. 2, eds. C.S. Cooper and P.L. Grover, Springer-Verlag, London, 1990, pp. 503.
79. A.C.Beach and R.C. Gupta, *Carcinogenesis*, 1992, **13**, 1053.
80. W.P. Watson. *Mutagenesis*, 1987, **2**, 319.
81. K. Randerath, M.V. Reddy and R.C. Gupta, *Proc. Natl. Acad. Sci. USA*, 1981, **78**, 6126.
82. W.L. Reichert, J.E. Stein, B. French, P. Goodwin, and U. Varanasi, *Carcinogenesis*, 1992, **13**, 1475.

83. E. Randerath, H.P. Agrawal, J.A. Weaver, C.B. Bordelon and K. Randerath, *Carcinogenesis*, 1985, **6**, 1117.
84. K. Randerath, E. Randerath, H.P. Agrawal,R.C. Gupta,M.E. Schurdak and M.V. Reddy, *Environmental Health Perspectives*, 1985, **62**, 57.
85. R.C. Gupta, *Cancer Res.*, 1985, **45**, 5656.
86. M.V. Reddy and K. Randerath, *Carcinogenesis*, 1986, 7, 1543.
87. D.H. Phillips, A. Hewer, C.N. Martin, R.C. Garner, and M.M. King, *Nature*, 1988, **36**, 790.
88. J. Cuzick, M.N. Routledge, D. Jenkins and R.C. Garner, *International Journal of Cancer*, 1990, **45**, 673.
89. K. Randerath and E. Randerath In *Human carcinogen exposure: biomonitoring and risk assessment*, eds. R.C. Garner, P.B. Farmer, and A.S Wright, Oxford University Press, Oxford, 1991, pp. 25.
90. K. Randerath, M.V. Reddy and R.M. Disher, *Carcinogenesis* 1986, 7, 1615.
91. K. Randerath, D. Li, B. Moorthy and E. Randerath *in Postlabelling methods for the detection of DNA adducts*, eds D.H. Phillips, M. Castegnaro and H. Bartsch, IARC, Lyon, 1993, pp 157.
92. M.V. Reddy, R.C. Gupta, E. Randerath and K. Randerath, *Carcinogenesis*, 1984, **5**, 231.
93. K. Hemminki, A. Forsti, M. Lofgren, D. Segerbäck, C. Vaca and P. Vodicka in *Postlabelling methods for detection of DNA adducts*, ed. D.H Phillips, M. Castegnaro, and H. Bartsch, IARC, Lyon, 1993, pp. 51.
94. V.L. Wilson, A.K. Basu, J.M. Essigmann, R.A. Smith and C.C. Harris, *Cancer Res.*, 1988, **48**, 2156.
95. W.P Watson and A.E. Crane, *Mutagenesis*, 1989, **4**, 75.
96. M.S.T. Steenwinkel, R. Roggeband, J.H.M. van Delft, and R. Baan in *Postlabelling Methods for Detection of DNA Adducts*, ed. D.H. Phillips, M. Castegnaro and H. Bartsch, IARC, Lyon, 1993, pp. 65.
97. A. H. S. Lau and W.H. Baird, *Carcinogenesis*, 1991, **12**, 885.
98. D.E.G. Shuker, M.-J. Durand and D. Molko, in *Postlabelling methods for the detection of DNA adducts* ed. D.H. Phillips, M., Castegnaro and H. Bartsch, IARC, Lyon, 1993 pp 157.
99. W.M. Baird, H.H.S. Lau, I. Schmerold and S.L. Coffing, In *Postlabelling methods for detection of DNA adducts*, eds. D.H. Phillips, M., Castegnaro and H. Bartsch, IARC, Lyon, 1993, pp. 217.
100. N.J. Gorelick *Mutat. Res.*, 1993, **288**, 5.
101. L. Möller, M. Ziesig and P. Vodicka, *Carcinogenesis*, 1993, **14**, 1343.
102. P.T. Strickland and J.M. Boyle, *Progress in Nucleic Acid Research and Molecular Biology*, 1984, **31**, 1.
103. R.M. Santella, X.Y. Yang, L.L. Hsieh and T.L. Young in *Mutation and the Environment*, Part C:Somatic and heritable mutation, adduction, and epidemiology ed. Vol. 340C, Wiley-Liss, Inc, New York, 1990, pp 247.
104. M.C. Poirier In *Human Carcinogen Exposure: Biomonitoring and Risk Assessment*, ed R.C. Garner, P.B. Farmer, and G.T. Steel, IRL Press, Oxford 1991, pp. 69.
105. R.A. Baan, O.B Zaalberg, A.J. Fichtinger-Schepman, M.A. Muysken-Schoen, M.J. Lansbergen and P.H.M. Lohman,*Environmental Health Perspectives*, 1985, **62**, 81.
106. R. Muller and M.F. Rajewsky, *Journal of Cancer Research and Clinical Oncology*, 1981, **102**, 99.
107. C.P. Wild, R. Saffhill and J.M. Boyle, *Carcinogenesis*, 1983, **4**, 1605.
108. R. Muller and M.F. Rajewsky, *Cancer Res,*, 1980, **40**, 887.
109. M.F. Rajewsky, R. Muller, J. Adamkiewicz and W. Drosdziok, *in Carcinogenesis fundamental mechanisms and environmental effect*, ed. B. Pullman, P.O.P Ts'o and H. Gelboin, Reidel, Dordrecht, 1980, pp207.
110. C.J. van der Laken, A.M. Hagenaars,G. Hermsen, E. Kriek, A.J. Kuipers, J. Nagel, E. Scherer, M. Welling, *Carcinogenesis*, 1982, **3**, 569.
111. A.A. Wani, R.E. Gibson-D'Ambrosio, S.M. D'Ambrosio, *Carcinogenesis*, 1984, 5, 1145.

112. R.A. Baan, M.J. Lansbergen,P.A.F. de Bruin, M.I. Willems, P.H.M. Lohman, *Mutat. Res.*, 1985, **150**, 23.
113. Guigues and Leng 1979
114. M.C. Poirier, S.H. Yuspa, I.B. Weinstein, S. Blobstein, *Nature*, 1977, **270**, 186.
115. P. Rio and M. Leng, *Biochimie*, 1980, **62**, 487.
116. M. Spodheim-Maurizot and M. Leng, *Carcinogenesis*, 1, 1980, 807.
117. M. Leng, and E. Sage, *FEBS Lett.*, 1978, **92**, 207.
118. E. Sage, P.P. Fuchs and M. Leng, *Biochemistry*, 1979, **18**, 1328.
119. M.C. Poirier, R. Santella, I.B. Weinstein, D. Grunberger, S.H. Yuspa, *Cancer Research*, 1980, **40**, 412.
120. R. M. Santella, C.D. Lin, W.L. Cleveland, I.B. Weinstein, *Carcinogenesis*, 1984, 5, 373.
121. H. Slor, H. Mizusawa, N. Neihart, T. Kakefuda, R.S. Day, and M. Bustin, *Cancer Res.*, 1981, **41**, 3111.
122. A.M.J. Fichtinger-Schepman, R.A. Baan, A. Luiten-Schuite, M. Van Dijk and P.M.H. Lohman, *Chemico-Biological Interactions*, 1985, **55**, 275.
123. B.Malfoy, B. Hartmann, J.P. Macquet and M. Leng, *Cancer Res.*, 1981, **41**, 4127.
124. M.C. Poirier, S.J. Lippard, L.A. Zwelling, H.M. Ushay, D. Kerrigan,C.C. Thill, R.M.
125. R.A. Santella, D. Grunberger, S.H. Yuspa, *Proceedings of the American Association for Cancer Research*, 1982, 79, 6443.
126. L.L. Hsieh, A.M. Jeffrey and R.M. Santella, *Carcinogenesis*, 1985, **6**, 1289.
127. R.M. Santella, N. Dharmaraja and R.L. Edelson *Nucleic Acids Res.*, 1985, 13, 2533.
128. A. Haugen, J.D. Groopman, I.C. Hsu, G.R. Goodrich, G.N. Wogan and C.C. Harris, *Proc. Natl. Acad. Sci. USA*, 1981, **78**, 7.
129. P.J. Hertzog, J.R. Lindsay-Smith and R.C Garner, *Carcinogenesis*, 1982, 3, 825.
130. L.L. Hsieh, S.W. Hsu, D.S. Chen and R.M. Santella, *Cancer Res*, 1988, **48**, 6328.
131. S.A. Leadon and P.C. Hanawalt, *Mutat. Res.*, 1983, **112**, 191.
132. G.J. West, I.W-L. West and J.F. Ward, *Radiation Research*, 1982, **90**,595.
133. R.D. Ley, *Cancer Res.*, 1983, **43**, 41.
134. P.T. Strickland and J.M. Boyle, *Photochem. Photobiol.*, 1981, **34**, 595.
135. M.J. Tilby, J.M. Styles and C.J. Dean, *Cancer Res.*, 1987, **47**, 1542.
136. F.J. van Schooten, E. Kriek, M.S.T. Steenwinkel, H. P.J.M. Noteborn, M.J.X. Hillebrand and F.E.V. Leeuwen *Carcinogenesis*, 1987, **8**, 1263.
137. R.M. Santella, A. Weston, F.P. Perera, G.E. Trivers, C.C. Harris, T.L. Young, D. Nguyen, B.M. Lee, and M.C. Poirier *Carcinogenesis*, 1988, **9**, 1265.
138. M.M. King, J. Cuzick, D. Jenkins, M.N. Routledge and R.C. Garner, *Mutat. Res.*, 1993, **292**, 113.
139. A.S. Wright, in *Human Carcinogen Exposure*, ed. R.C. Garner, P.B. Farmer, G.T. Steel and A.S. Wright, Oxford University Press, Oxford, 1992, pp. 3.
140. L. den Engelse, J. Van Benthem and E. Scherer, *Mutat. Res.*, 1990, **233**, 265.
141. H. Slor, N. Mizusawa, T. Nechart, R. Kakefuda, R.S. Day and M. Bustin, *Cancer Res.*, 1981, **41**, 3111.
142. F.J. van Schooten, M.J.X. Hillebrand, E. Scherer , L. den Engelse and E. Kriek, *Carcinogenesis*, 1991, **12** .
143. C-.J. Chen, Y.-J. Zhang, S.-N Lu, and R.M. Santella, *Hepatology*, 1992, **16**, 1150.
144. R.S. Paules, M.C. Poirier, M.J. Mass, S.H. Yuspa and D.G. Kaufman, *Carcinogenesis*, 1985, **6**, 193.
145. B. Tierney, A. Benson and R.C. Garner, *Journal of the National Cancer Institute*, 1986, 77, 261.
146. M.M. King, C. Schell and R.C. Garner in *Human Carcinogen Exposure*, ed. R.C.Garner, P.B. Farmer, G.T. Steel and A.S. Wright, Oxford University Press, Oxford, 1992, pp 293.
147. J.D. Groopman, A.J. Hall, H. Whittle, G.J. Hudson, G.N. Wogan, R. Montesano and C.P. Wild). *Cancer Epidemiology, Biomarkers & Prevention*, 1992, **1**, 221.
148. E.C. Miller, *Cancer Res.*, 1951, **11**, 100.

149. D.K. Manchester, A Weston, J.S. Choi, G.E. Trivers, P. Fennessey, E. Quintana, P.B. Farmer, D.L. Mann and C.C. Harris, *Proc. Natl. Acad. Sci. USA*, 1988, **85**, 9243.
150. H. Autrup, K.A. Bradley, A.K.M. Shamsuddin, J. Wakhisi and A. Wasunna, *Carcinogenesis*, 1983, **4**, 1193.
151. V.M. Weaver and J.D. Groopman, *Cancer Epidemiology, Biomarkers & Prevention*, 1994, **3**, 669.
152. A. Haugen, G. Becher, C. Benestad, K. Vahakangas, G.E. Trivers, M.J. Newman and C.C. Harris, *Cancer Res.*, 1986, **46**, 4178.
153. C.C. Harris, A. Weston, J.C. Willey, G. Trivers, D.L. Mann, *Environmental Health Perspectives*, 1987, **75**, 109.
154. P.B. Farmer, D.E.G. Shuker, and I. Bird, *Carcinogenesis*, 1986, **7**, 49.
155. E. Bailey, P.B. Farmer and D.E.G. Shuker, *Arch. Toxicol.*, 1987, 60, 187.
156. P.S. Skipper and S. Naylor in *Human Carcinogen Exposure*, ed. R.C. Garner, P.B. Farmer,G.T.Steel and A.S. Wright, Oxford University Press, Oxford, 1992, pp61.
157. R.W. Giese, in *Human carcinogen exposure: biomonitoring and risk assessment*, ed. R.C. Garner, P.B. Farmer, G.T. Steel and A.S. Wright, IRL Press, Oxford, 1991, pp 247.
158. J. Cadet, M.-F. Incardona, F. Odin, D. Molko, J.-F. Mouret, M. Polverelli, H. Faure, V. Ducros, M. Tripier and A. Favier A, in *Postlabelling methods for detection of DNA adducts*, ed. D.H. Phillips, M. Castegnaro and H. Bartsch, IARC, Lyon, 1993, pp 271.
159. D.K. Manchester,A. Weston,J.S. Choi,G.E. Trivers,P. Fennessey, E. Quintana,P.B. Farmer, D.L. Mann, C.C. Harris, *Proc. Natl. Acad. Sci. USA*, 1988, **85**, 9243.
160. A.K. Chaudhary, M. Nokubo, G.R. Reddy, S.N. Yeola, J.D. Morrow, I.A. Blair and L.J. Marnett, *Science*, 1994, **265**, 1580.
161. R. Barak, A. Vincze, P. Bel, S.P. Dutta and G.B. Chedda, *Chemico-Biological Interactions*, 1993, **86**, 29.
162. N. Potier, A.V. Dorsselaer, Y. Cordier, O. Roch and R. Bischoff, *Nucleic Acids Res.*, 1994, **22**, 3895-3903.
163. J.R. Chapman, *Practical organic mass spectrometry*, Wiley, Chichester, 1993.
164. J.J. Dino, C.R. Guenat, K.B. Tomer and D.G. Kaufman *Rapid Commun. Mass Spec.*, 1987, **1**, 69.
165. K.W. Turteltaub, J.S. Felton, B.L. Gledhill, J.S. Vogel, J. R. Southon, M.W. Caffee, R.C. Finkel, D.E. Nelson, I.D. Proctor and J.C. Davis, *Proc. Natl. Acad. Sci. USA*, 1990, **87**, 5288.
166. K.W. Turteltaub, J.S. Vogel, C.E. Frantz and E. Fultz, in *Postlabelling methods for detection of DNA adducts*, ed. D.H. Phillips, M. Castegnaro and H. Bartsch, IARC, Lyon, 1993, pp 293.
167. V.L. Souliotis, and S.A. Kyrtopoulos, *Cancer Res.*, 1989, **49**, 6997.
168. V.L. Souliotis, S. Kaila, V.A. Boussiotis, G.A. Pangalis and S.A. Kyrtopoulos, *Cancer Res.*, 1990, **50**, 2759.
169. S.A. Kyrtopoulos, P. Ampatzi, N. Davaris, N. Haritopoulos and B. Golematis, *Carcinogenesis*, 1990, **11**, 431-436.
170. B. Vogelstein and K.W. Kinzler, *Nature*, 1992, **355**, 209.
171. B. Vogelstein and K.W. Kinzler, *Trends in Genetics*, 1993, **9**, 138.
172. V.A. Bohr, D.H. Phillips and P.C. Hanawalt, Heterogenous DNA damage and repair in the mammalian genome. *Cancer Res.*, 1987, **47**, 6426.
173. V.A. Bohr and K. Wassermann, *Trends in Pharmacological Science*, 1988, **13**, 429.
174. V.A. Bohr, *Carcinogenesis* 1991, **12**, 1983.
175. V.A. Bohr, C.A. Smith, D.S. Okumoto, and P.C. Hanawalt, *Cell*, 1985, **40**, 359.
176. K. Nose and O. Nikaido, *Biochim. Biophys. Acta.*, 1974, **781**, 273.
177. D.C. Thomas, A.G. Morton, V.A. Bohr, and A. Sancar *Proc. Natl. Acad. Sci. USA*, 1988, **85**, 3723.
178. G.P. Pfeifer, R. Drouin and G.P. Holmquist, *Mutat. Res.*, 1993, **288**, 39.
179. D.P. Kalinowski, S. Illenye and B. Van Houten, *Nucleic Acids Res.*, 1992, **20**, 3485.

180. U. Landegren, R. Kaiser, J. Sanders and L.Hood, *Science*, 1988, **241**, 1077.
181. C.S. Lee, G.P. Pfeifer and N.W.Gibson *Cancer Res.*,1994, **54**, 1622.
182. S. Tornaletti, D. Rozek and G.P. Pfeifer, *Oncogene*, 1993, **8**, 2051.
183. G.P. Pfeifer, R. Drouin, A.D. Riggs and G.P. Holmquist, *Proc. Natl. Acad. Sci. USA* 1991, **88**, 1374.
184. D.H. Phillips, B. Schoket, A. Hewer, E. Bailey, S. Kostic and I. Vincze, *Journal of the National Cancer Institute.*, 1990, **46**, 569.
185. P.G. Foiles, L.M. Miglietta, A.M. Quart, E. Quart, G.C. Kabat and S.S. Hecht, *Carcinogenesis*, 1989, **10**, 1429.
186. R.B. Everson, E. Randerath, R.M. Santella, R.C. Cefalo, T.A. Avitts and K. Randerath, *Science*, 1986, **231**, 54.
187. F.J. van Schooten, M.J.X. Hillebrand, F.E van Leeuwen, N. van Zandwijk, H.M. Jansen, L. den Engelse and E.Kriek, *Carcinogenesis*, 1992,13, 987.
188. F.P. Perera, K. Hemminki, E. Gryzbowska, G. Motykiewicz, J. Michalska, R.M. Santella, T.-L. Young, C. Dickey, P. Brandt-Rauf, I. DeVivo, W. Blaner, W.-Y. Tsai and M. Chorazy, *Nature*, 1992, **360**, 256.
189. D.E.G. Shuker and P.B. Farmer, *Chemical Research in toxicology.*, 1992, **5**, 450.
190. J.D. Groopman, J. Zhu, P.R. Donahue, A. Pikul, L.-S. Zhang., J.-S. Chen and G.N. Wogan, *Cancer Res.*, 1992, **52**, 45.
191. R.A. Bennett, J.M. Essigmann and G.N. Wogan, *Cancer Res.*, 1981, **41**, 650.
192. R.K. Ross, J-M. Yuan, M.C. Yu, G.N. Wogan, G-S. Qian, J-T. Tu, J.D. Groopman, Y-T. Gao, B.E. Henderson, *The Lancet*, 1992, **339**, 943.
193. M.S. Greenblatt, W.P. Bennett, M. Hollstein and C.C. Harris, *Cancer Res.*, 1994, **54**, 4855.
194. F.H. Sobels, *in Progress in Mutat. Res., vol 3,* ed. K.C. Bora, G.R. Douglas and E.R. Nestman, Elsevier Amsterdam, 1982, pp323.
195. L. Ehrenberg, E. Moustacchi, S. Osterman-Golkar and G. Ekman, *Mutat.Res.* 1983, **23**, 121.
196. M. Tornqvist, D. Segerback and L. Ehrenberg in *Human carcinogen exposure: biomonitoring and risk assessment,* eds. R.C. Garner, P.B. Farmer, and A.S Wright, Oxford University Press, Oxford, 1991, pp.141.
197. H.D. Madhani, V.A. Bohr and P.C. Hanawalt, *Cell*, 1986, **45**, 417.
198. I. Mellon and P.C. Hanawalt, *Nature* 1989, **342**, 95.

The Use of Biomarkers in Food Chemical Risk Assessment

D. R. Tennant

FOOD SAFETY DIRECTORATE, MINISTRY OF AGRICULTURE, FISHERIES AND FOOD, ERGON HOUSE, C/O NOBEL HOUSE, 17 SMITH SQUARE, LONDON SW1P 3JR, UK

1 INTRODUCTION

Biomarkers have been identified as one of the most promising developments in chemical risk assessment. However, as the term 'biomarkers' gets used more, so the diversity of its meanings seems to expand. This paper will try to define this term but before doing so it is necessary first to understand more about biomarkers: - Why do we need them? What can we do with them? How should we measure them and what do they mean? This chapter represents the views of the Symposium organising committee and attempts to answer these questions.

2. THE NEED FOR BIOMARKERS

Current techniques for monitoring the safety of chemicals in food supply rely on measuring concentrations in foods and then estimating likely intakes from data on food consumption patterns. The intake estimate is then compared with an estimate of an acceptable intake which is derived by extrapolation from animal studies. There are therefore many stages where uncertainties can be introduced in the current risk assessment process (Figure 1). Each stage in this process involves expert judgement, modelling and estimation. The product is the result of a lengthy series of calculations. The risk assessment is therefore based on a calculation not an observation.

To overcome each source of uncertainty, safety factors or conservative assumptions are used to ensure that the risk assessment always errs on the side of safety. Whilst this gives a large margin of consumer safety, it is a very inefficient method for decision making.

Biomarkers offer a means of short-cutting this process so that measurements can be made on the individual at risk. This is the first potential benefit from the use of biomarkers:

~ *Biomarkers are more direct and more accurate.*

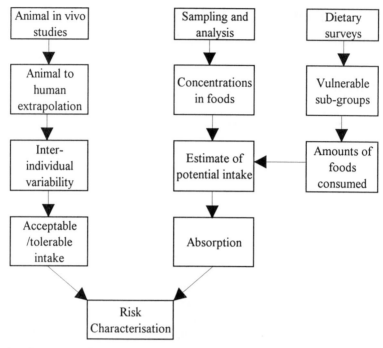

Figure 1. *Conventional approach to food chemical risk assessment.*

Conventional risk assessment also measures the effects of chemicals in isolation and often at high doses. It does not take into account the influence of other factors which might affect the final outcome, in particular other constituents of the diet. These other constituents could affect, for example:

> absorption
> pharmacodynamics
> metabolism (through enzyme induction, etc.)
> excretion
> individual susceptibility
> beneficial effects of other constituents of the diet
> synergistic/agonistic effects of other constituents of the diet
> other sources of exposure

Conventional approaches to risk assessment therefore lack relevance to human exposures and this is the second potential benefit from using biomarkers:

> ~ *Biomarkers produce more relevant data.*

Biomarkers can provide either direct measures of human effects or have the potential to be correlated directly with human effects. If they are accurate, relevant and health-related then they offer a third potential benefit:

> ~ *Biomarkers can produce direct evidence of human risk.*

Whilst the biomarker approach offers the prospect of more accurate and relevant risk assessment it cannot remove all sources of uncertainty. Indeed biomarker techniques introduce new uncertainties which must be carefully assessed before embarking on any new biomarker approach.

3 POTENTIAL USES FOR BIOMARKERS

Many potential applications for biomarkers have been discussed during this symposium (Table 1). Some of these are available for immediate application whereas the development of others is a long way off. It is important to note the diversity of potential applications. This is another strength of the biomarker approach - its versatility.

Table 1. *Potential uses for biomarkers*

1.	Setting standards related to health effects
2.	Assessing exposure in terms of toxicologically relevant dose
3.	Hazard prioritisation
4.	Validation of conventional intake estimates
5.	Indicating levels of risk
6.	Measurement of internal (target site) dose
7.	Assessing bioavailability
8.	Estimating risk/exposure to individuals
9.	Estimating risk/exposure to populations
10.	Predicting toxicological consequences
11.	Low dose extrapolation
12.	Identification of human metabolic pathways
13.	Quantification of inter-individual variability
14.	Identifying populations at risk
15.	Estimating food consumption
16.	Validation of 24 hour urine collections
17.	Evaluation of toxins in whole foods or entire diet
18.	Exposure source tracing
19.	Internal dosimetry (microcapsules)
20.	Retrospective exposure monitoring
21.	Exposure/intake of beneficial agents in diet
22.	Measurement of endogenous toxins
23.	Exposure to activated intermediates
24.	Measurements of oxidative stress
25.	Measurement of chemoprotective enzymes
26.	Monitoring role of diet in stimulating defences
27.	Investigating dose-response relationships

Whilst some of the uses listed in Table 1 might overlap there are probably many other uses not mentioned and possibly more yet undiscovered.

4   HOW SHOULD BIOMARKERS BE MEASURED?

Biomarkers present a new and exiting research field.   However, there is a need to consider carefully the approach to any biomarker study if the results produced are to be meaningful.   It is important to resist the temptation to start collecting data before methods have been validated and the particular biomarker system is fully understood.

4.1.   Pharmacokinetics. Preliminary studies of the literature and, if necessary, in the laboratory will need to be undertaken to establish the approximate kinetics of absorption, distribution, metabolism and excretion.   This is essential if an appropriate sampling regime is to be established.   For example, if the half time is of a few hours duration it could be pointless to collect an early morning blood or urine sample and levels might be affected by a recent meal.   On the other hand, there might be a time delay before any biological effects can be detected.

4.2   Relevant Biomarker. It is important that the marker should be relevant to the aim of the study.   Either the parent compound, a metabolite or some biological response could be measured.   In some cases more than one biomarker might be appropriate.

4.3   Interactions with the Diet. Checks must be made to find out how other components of the diet could affect absorption, distribution, metabolism or excretion of the chemical.   Can other factors affect the expression of any biological response?

4.4.   Validated Analytical Method. Analytical methods should be checked using internal and external quality assurance procedures to ensure that results are accurate. The performance of the method should be assessed and recorded.

4.5   Dose-Response. It is important to understand the dose-response relationship - either between the marker and dose or the marker and anticipated effect, depending on how the biomarker is being used.   Without this information it will be impossible to interpret the results.

5.   INTERPRETATION OF RESULTS

There is a great danger in collecting biomarker data when the significance of the results is not understood.   For example, data on DNA-adducts of certain environmental chemicals might be relatively easy to obtain.   However, the presence of adducts does not necessarily correlate with cancer risk.   On the other hand it is impossible to say that there is no risk.   This situation could be misused to generate public concern and create pressure for regulation when there is no evidence that any significant risk really exists.

The research priority is then for studies into the interpretation of biomarkers.   In particular, studies which are linked to epidemiological or mechanistic studies.   Research into biomarkers needs to be linked to conventional *in vivo* and novel *in vitro* toxicological approaches if it is to generate data which can be interpreted.   Biomarkers can complement the *in vivo / in vitro* parallelogram paradigm by adding the opportunity to make direct and comparable measurements in human and animal studies (Figure 2).

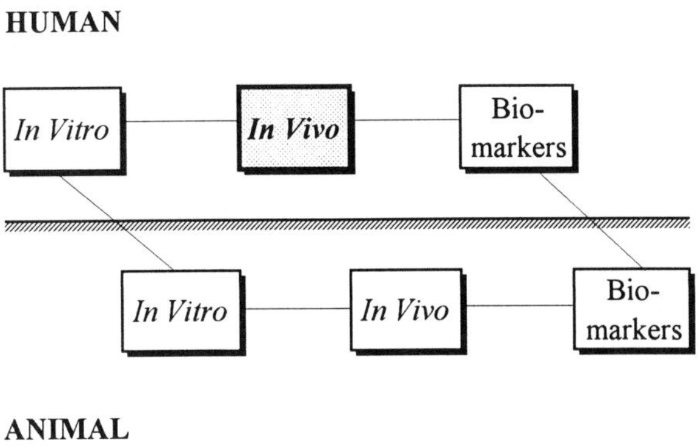

HUMAN

ANIMAL

Figure 2. *'Parallelogram' risk assessment paradigm.*

6.   WHAT IS THE FUTURE FOR BIOMARKERS?

In the short term it is likely that biomarkers will be used successfully as a method for overcoming the short-comings of conventional exposure assessments but only for those chemicals where the pharmacokinetics are relatively well understood and there is a direct relationship between absorption and excretion. This applies particularly to those chemicals which are fully absorbed but excreted completely and without metabolism. Here urinary excretion can provide a useful and valid method for assessing intake.

The use of biomarkers to predict risks reliably will be delayed until mechanisms of toxicity are studied and the relationships between biomarkers such as DNA adducts and risks better understood. There is some hope for the shorter term though: If more was known about the apparent background levels of adducts then it might be possible to tell whether adducts of any specific chemical were contributing significant amounts. This would be valuable information for regulators who need to make decisions about the need for controls or for providing advice and information to consumers.

7.   WHAT <u>ARE</u> BIOMARKERS?

Definitions of terms often take time to evolve. However, there seems to be a growing consensus that biomarkers can be sub-divided into three main groups (Table 2). From these groups of biomarkers a tentative definition can be proposed:

> ~ **A biomarker is any chemical which can be measured with minimal invasion,** *in vivo*, **in a body tissue or fluid, the concentration of which can be related directly to a factor of concern such as intake, dose, biological response, susceptibility, harmful effect, beneficial effect or risk.**

Table 2. *Types of Biomarkers.*

Exposure Biomarkers	Indicate the exposure or dose (recent, cumulative, internal, target site, etc.) of parent compound, metabolite, reactive intermediate or adduct.
Effect/Response Biomarkers	Indicate the biological response - either as an indirect measure of exposure or as a measure of a toxicologically relevant biological reaction.
Susceptibility Biomarkers	Predict the exposure/effect or dose/response relationship for an individual or specific sub-population; identify phenotypes/genotypes.

A combination of biomarker approaches may be necessary if all factors which can affect risks are to be taken into account. Once a biomarker, or suite of biomarkers, has been validated and the results can be interpreted, then it can be used routinely in 'biological monitoring' programmes.

8. CONCLUSIONS

The relationship between food and health is extremely complex. Biomarkers offer some exciting opportunities to overcome some of these complexities and to provide more direct, accurate and relevant methods for use in risk assessment. However, care must be taken in planning studies to ensure that results will be meaningful and not merely contribute to the confusion.

At the moment biomarkers research tends to be focused on carcinogens. This may be because there is already a lead in the understanding of mechanisms of toxicity in this area. However, there is a need to diversify the biomarkers approach to include other end-points such as feto-toxicity and immunotoxicity.

It is also important to remember that not all dietary constituents represent hazards. Many chemicals in foods are beneficial and some can mitigate the effects of toxic chemicals. Any new biomarker-based approach to risk assessment must take beneficial effects into account alongside harmful ones.

In the short-term biomarkers offer the opportunity to resolve many of the short-comings of the current indirect, model-based approaches to toxicology and risk assessment. This should allow a refinement of these approaches and support better decision-making.

The longer term offers more exciting prospects. Biomarkers are capable of measuring integrated doses to individuals and, more importantly, the net biological effect of exposure taking all exogenous and endogenous factors into account and combining harmful and beneficial influences. This could allow the risk assessment of whole foods or the entire diet and opens up a new perspective in risk assessment.

Finally, risk assessment of food chemicals is now moving away from national fora into an international arena - particularly in the EU. It is therefore necessary to develop biomarkers within a European risk assessment framework. This will need much collaboration so that methods are applicable and relevant in all member states and a European consensus on the use and interpretation of biomarker data can evolve.

Subject Index